Bayesian Methods for Ecology

The interest in using Bayesian methods in ecology is increasing, but most
ecologists do not know how to carry out the required analyses. This book bridges
that gap. It describes Bayesian approaches to analysing averages, frequencies,
regression, correlation and analysis of variance in ecology. The book also
incorporates case studies to demonstrate mark-recapture analysis, development
of population models and the use of subjective judgement. The advantages of
Bayesian methods, including the incorporation of any relevant prior information
and the ability to assess the evidence in favour of competing hypotheses, are
also described here. The analyses described in this book use the freely available
software WinBUGS, and there is an accompanying website (http://arcue.botany.
unimelb.edu.au/bayes.html) containing the data files and WinBUGS codes that
are used in the book. The Bayesian methods described here will be of use
to ecologists from the level of upper undergraduate and above.

MICHAEL A. MCCARTHY is Senior Ecologist at the Royal Botanical
Gardens, Melbourne and Senior Fellow in the School of Botany at the
University of Melbourne.

Bayesian Methods for Ecology

MICHAEL A. McCARTHY

CAMBRIDGE
UNIVERSITY PRESS

CAMBRIDGE UNIVERSITY PRESS
Cambridge, New York, Melbourne, Madrid, Cape Town, Singapore, São Paulo,
Delhi, Dubai, Tokyo, Mexico City

Cambridge University Press
The Edinburgh Building, Cambridge CB2 8RU, UK

Published in the United States of America by Cambridge University Press, New York

www.cambridge.org
Information on in this title: www.cambridge.org/9780521850575

First published 2007
4th printing 2011

Printed in the United Kingdom at the University Press, Cambridge

A catalogue record for this publication is available from the British Library

Library of Congress Cataloging-in-Publication Data

McCarthy, Michael A., 1968-
Bayesian methods for ecology / Michael A. McCarthy.
p. cm.
Includes bibliographical references and index.
ISBN-10: 0-521-85057-6
ISBN-13: 978-0-521-85057-5
ISBN-10: 0-521-61559-3 (pbk.)
ISBN-13: 978-0-521-61559-4 (pbk.)
1. Ecology–Research–Statistical methods. 2. Bayesian statistical decision
theory. I. Title.

QH541.2.M38 2007
577.072′4–dc22   2006102405

ISBN 978-0-521-85057-5 Hardback
ISBN 978-0-521-61559-4 Paperback

To Kirsten and Owen

Contents

Preface

I have three vivid memories about learning statistics as an undergraduate that all involve misconceptions. Firstly, I remember my lecturer telling me that, after obtaining a result that was not statistically significant, I should conclude that timber harvesting did not have an effect (on what, I cannot remember). While the logic was flawed, I have since realized that it is a misconception shared by many ecologists.

My second memory is of reading about Bayesian analyses in journal articles. I wondered what Bayesian methods were, how they differed from the statistical approaches I had been taught (frequentist methods such as null hypothesis testing and construction of confidence intervals), and why I had never heard of them before. On reading the articles, I concluded that Bayesian methods must be hard to do. It turns out that I was incorrect again.

My third memory is that statistics was boring. I was wrong again. I was reasonably good at the mathematics involved, but it was not until I started doing my own data analyses during my Ph.D. that I saw the benefits of using statistics. I began to learn about different ways to do statistics (e.g. likelihood-based methods), and also re-learnt some old topics (e.g. realizing the importance of and learning how to calculate statistical power). For me, statistics and probability continue to be a world of learning.

This book represents a stage in my journey through the world of statistics. It is born out of a frustration with how conventional statistical methods are misused in ecology on a routine basis, and a belief that Bayesian methods are relevant and useful. I hope this book convinces readers of the value of Bayesian methods and helps them learn Bayesian methods more quickly than me.

Approximately five years ago I used null hypothesis significance testing to evaluate the predictions of some models of population viability. An astute reviewer questioned this approach because the models were surely known to be wrong a priori. The reviewer provided a glorious list of quotes that attacked null hypothesis significance testing (not unlike the quotes in Chapter 2). I started thinking about alternatives, leading me to Hilborn and Mangel's (1997) *The Ecological Detective*, and beyond.

The Ecological Detective (Hilborn and Mangel, 1997) is one of the best books available to ecologists for learning about Bayesian methods. However, ecologists wishing to use the suggested methods need at least some skills in computer programming. I intend my book to provide a bridge between a desire to conduct Bayesian analyses and the book by Hilborn and Mangel (1997). WinBUGS code for the analyses in this book is available from http://arcue.botany.unimelb.edu.au/bayes.html.

The bridge is built by using the freely available program WinBUGS (Spiegelhalter *et al.*, 2005; Appendix A) to conduct the vast majority of analyses in this book. I try to start gently, illustrating the most basic analyses, before giving some more complicated examples. More experienced users will find some analyses trivial, and novices may find some examples impenetrable. The aim is to provide a sufficient diversity of examples that the reader will be able to learn how to construct their own statistical models and conduct their own analyses.

This book is not necessarily designed to be read from cover to cover. Read Chapters 1 and 2 if you wish to know more about the differences between Bayesian and frequentist methods. If you just want to learn how to conduct Bayesian analyses, start with Chapter 1, Appendix A, and then move to Chapter 3 or whichever topic is most relevant. As you become more familiar with Bayesian methods, the entire content of the book will become more accessible.

I have many people to thank for their help while writing this book. Ralph Mac Nally and Alan Crowden's suggestion to write this book started the ball rolling. Brendan Wintle has been extremely important as a colleague, a source of advice and insights, and a sounding board for ideas. Kirsten Parris, David Lindenmayer, Jane Elith, Pip Masters, Linda Broome, Tara Martin, Mark McDonnell, Michael Harper, Brendan Wintle, Amy Hahs, Rodney van der Ree and many others have provided data for analysis over the years. I would have learnt very little without them.

This book owes much to the availability of WinBUGS, and I thank the team that developed the software. In particular, David Spiegelhalter,

Andrew Thomas and Bob O'Hara have answered questions and investigated software bugs from time to time.

Hugh Possingham, Mark Burgman, David Lindenmayer and Mark McDonnell have provided opportunities for me to conduct research into risk assessment, and the use of Bayesian methods. They have been influential, as have Fiona Fidler, Neil Thomason and my colleagues listed above. Various funding agencies have supported my research, with the Australian Research Council and The Baker Foundation being particularly generous.

I'd like to thank my parents David and Sue for fostering my interests in mathematics and ecology, and Dr John Gault for his enthusiastic interest in Bayesian issues of a medical nature.

Finally, thank you to everyone who has provided comments, in particular Mark Burgman, Geoff Cumming, Aaron Ellison, Ralph Mac Nally, Kirsten Parris, Gerry Quinn and Julia Stammers, who read most if not all of earlier drafts. Peter Baxter, Barry Brook, Ryan Chisholm, Gareth Davies, Lou Elliott, Fiona Fidler, Bill Langford, Terry Walshe and Nick Williams also provided helpful comments on various sections.

1

Introduction

There is a revolution brewing in ecology. Granted, it is a gentle and slow revolution, but there is growing dissatisfaction with the statistical methods that have been most commonly taught and used in ecology (Hilborn and Mangel, 1997; Wade, 2000; Clark, 2005).[1] One aspect of this revolution is the increasing interest in Bayesian statistics (Fig. 1.1). This book aims to foster the revolution by making Bayesian statistics more accessible to every ecologist.

Ecology is the scientific study of the distribution and abundance of biological organisms, and how their interactions with each other and the environment influence their distribution and abundance (Begon *et al.*, 2005). The discipline depends on the measurement of variables and analysis of relationships between them. Because of the size and complexity of ecological systems, ecological data are almost invariably subject to error. Ecologists use statistical methods to distinguish true responses from error. Statistical methods make the interpretation of data transparent and repeatable, so they play an extremely important role in ecology.

The Bayesian approach is one of a number of ways in which ecologists use data to make inferences about nature. The different approaches are underpinned by fundamentally different philosophies and logic. The appropriateness of different statistical approaches has been fiercely debated in numerous disciplines but ecologists are only now becoming aware of this controversy. This occurs at least in part because the majority of statistical books read by ecologists propound conventional

---

[1] The conventional statistical methods are known as frequentist statistics and include null hypothesis significance testing (NHST) and construction of confidence intervals. NHST attracts the most criticism. See Chapter 2 for more details of these methods.

1

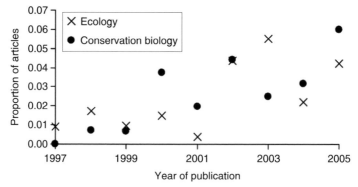

Fig. 1.1 The proportion of articles in the journals *Ecology* and *Conservation Biology* that refer to 'Bayes' or 'Bayesian'.

statistics, ignore criticisms of these methods and do not acknowledge that there are alternatives (Fowler *et al.*, 1998; Sokal and Rohlf, 1995; Underwood, 1997; Zar, 1999). Those that do address the controversy usually aim to change the status quo (Hilborn and Mangel, 1997; Burnham and Anderson, 2002), although there are exceptions (Quinn and Keough, 2002; Gotelli and Ellison, 2004).

The Bayesian approach is used relatively rarely (Fig. 1.1), so why should it interest ecologists? There are several reasons but two are particularly relevant ones. Firstly, Bayesian methods are fully consistent with mathematical logic, while conventional statistics are only logical when making probabilistic statements about data, not hypotheses (Cox, 1946; Berger and Berry, 1988; Jaynes, 2003). Bayesian methods can be used to make probabilistic predictions about the state of the world, while conventional statistics are restricted to statements about long-run averages obtained from hypothetical replicates of sampled data.

Secondly, relevant prior information can be incorporated naturally into Bayesian analyses by specifying the appropriate prior probabilities for the parameters. In contrast, conventional statistical methods are forced to ignore any relevant information other than that contained in the data. Difficulties with Bayesian methods and other benefits are discussed more fully in Chapter 2 and throughout this book.

Bayesian statistics are founded on the work of the Reverend Thomas Bayes, who lived and died in eighteenth century England (Box 1.1). Bayesian methods explicitly recognize and combine four

---

Box 1.1
The Reverend Thomas Bayes, FRS

Very little is known about Thomas Bayes. The portrait above
(O'Donnell, 1936) may be of Bayes, but no other portraits
are known (Bellhouse, 2004). Even the year (1701 or 1702) and
place of his birth (London or Hertfordshire, England) are
uncertain (Dale, 1999). There are few records to indicate the nature
of his early schooling, but he is known to have studied divinity
and mathematics at the University of Edinburgh. He was ordained
as a Presbyterian minister by 1728. He was elected as a Fellow of
the Royal Society in 1742 but it was not until after his death in
1761 that his most famous contribution, his essay in the
Philosophical Transactions of the Royal Society of London,
was published (Bayes, 1763). In that essay, Bayes described his
theory of probability and presented what is now known as
Bayes' rule (or Bayes' theorem), establishing the basis of
Bayesian statistics.

---

components of knowledge. Prior knowledge and new data are combined
using a model to produce posterior knowledge.[2] These four components
may be represented as:

$$\text{prior} + \text{data} \xrightarrow{\text{model}} \text{posterior}$$

It is common in everyday life to combine prior information and
new data to update knowledge. We might hear a weather forecast that the
chance of rain is small. However, if we stepped outside and saw dark

---

[2] Prior and posterior refer to before and after considering the data.

clouds looming above us, most people would think that the risk of rain was higher than previously believed. In contrast, our expectation of a fine day would be reinforced by a sunny sky. Thus, both the prior information (the weather forecast) and the data (the current state of the weather) influence our newly updated belief in the prospects of rain.

Our updated belief in the chance of rain (the posterior) will depend on the relative weight we place on the prior information compared to the new data and the magnitude of the difference between the two pieces of information. In this case the 'model' is contained within our understanding of the weather. Our thought processes combine the prior information, data, and model to update our belief that it will rain. Bayesian statistics provide a logically consistent, objective and repeatable method for combining prior information with data to produce the posterior, rather than the subjective judgement that most people would use when stepping outside.

Before considering the benefits and limitations of Bayesian methods and its alternatives in Chapter 2, I will illustrate the use of the different statistical approaches with two examples. These highlight how Bayesian methods provide answers to the kinds of questions that ecologists ask, and how they can usefully incorporate prior information.

Example 1: Logic in determining the presence or absence of a species

Consider an ecologist who surveys ponds in a city for frogs. On her first visit to a pond, she searches the edge and listens for frog calls over a 20-minute period. The southern brown tree frog (*Litoria ewingii*) is the most common species in her study area, but it is not found on this particular visit (Fig. 1.2). However, the researcher would not be particularly surprised that the species was not detected because she knows from experience that when surveying ponds, southern brown tree frogs are detected on only 80% of visits when they are in fact present. Given this information, what can she conclude about whether the southern brown tree frog is present at the site or not?

The question about the presence of a species is a simple example of those asked by ecologists. We assume that there is a particular true state of nature and we hope to use scientific methods to determine a reasonable approximation of the truth. However, the probability that a species is

Example 1 5

Fig. 1.2 The southern brown tree frog *Litoria ewingii*, a common species in the ponds of Melbourne, Victoria. Photograph by Nick Clemann.

present at a site is rarely calculated by ecologists, although it should be a fundamental part of any field study that depends on knowing where a species does and does not occur. This probability is not calculated partly because the statistical methods used by most ecologists are not well-suited to this question. I will examine three different approaches to answering this question and demonstrate that a satisfactory answer requires Bayesian methods.

Frequentist approaches

Conventional approaches to data analysis in ecology estimate the likelihood of observing the data (and more extreme data in the case of null hypothesis testing). These approaches are referred to as frequentist methods because they are based on the expected frequency that such data would be observed if the same procedure of data collection and analysis was implemented many times. Frequentist methods focus on the frequency with which the observed data are likely to be obtained from hypothetical replicates of sampling.

There are numerous types of frequentist statistics that are used in ecology, including null hypothesis significance testing and information-theoretic methods. These are applied below to the question about whether southern brown tree frogs are present at the pond.

Null hypothesis significance testing

The first statistical approach to answering the question is null hypothesis significance testing. The null hypothesis for this first case might be that the southern brown tree frog is absent from the site. The researcher then seeks to disprove the null hypothesis with the collection of data. The single piece of data in this case is that the frog was not detected. The researcher then asks: 'What is the probability of obtaining this result if the null hypothesis were true?'[3] This probability is the *p*-value of the significance test. If the *p*-value is sufficiently small (conventionally if less than 0.05), it means that the data (or more extreme data) would be unlikely to occur if the null hypothesis is true. If the *p*-value is small, then we assume that the data are inconsistent with the null hypothesis, which is then rejected in favour of the alternative.

In the case of the frog survey, the *p*-value is equal to 1.0. This is calculated as the probability that we would fail to record the frog (i.e. obtain the observed data) if it is absent (i.e. if the null hypothesis is true). The high *p*-value means that the researcher fails to reject the null hypothesis that the frog is absent.

The other possible null hypothesis is that the frog is present at the site. In this case, the probability of obtaining the data is equal to 0.2 (one minus the probability of detecting the species if present) given that the null hypothesis is true. Thus, the *p*-value is 0.2, and using a conventional cut-off of 0.05, the researcher would have a non-significant result. The researcher would fail to reject the null hypothesis that the southern brown tree frog was present.

It is surprising (to some people) that the two different null hypotheses can produce different results. The conclusion about whether the species is present or absent simply depends on which null hypothesis we choose. The source of this surprise is our failure to consider statistical power, which I will return to in Chapter 2.

Another possible source of surprise is that the *p*-value does not necessarily provide a reliable indicator of the support for the null hypotheses. For example, the *p*-value is equal to 1.0 for the null hypothesis that the frog is absent. This is the largest possible *p*-value, but it is still not proof that the null hypothesis is true. If we continued to return to the

---

[3] In actual fact, a null hypothesis significance test asks what is the probability of obtaining the data *or a more extreme result*. However, in this case, a more extreme result is not possible; it is not possible to fail to detect the frog more than once with one visit, so the *p*-value is simply the probability of observing the data.

Example 1 7

same pond and failed to find the frog, the p-value would remain equal to 1.0, insensitive to the accumulation of evidence that the frog is absent. This apparent discrepancy occurs because frequentist methods in general and p-values in particular do not provide direct statements about the reliability of hypotheses (Berger and Sellke, 1987; Berger and Berry, 1988). They provide direct information about the frequency of occurrence of data, which only gives indirect support for or against the hypotheses. In this way, frequentist methods are only partially consistent with mathematical logic, being confined to statements about data but not directly about hypotheses (Berger and Sellke, 1987; Jaynes, 2003).

Information theoretic methods

An information theoretic approach based on 'likelihood' is an alternative frequentist method to null hypothesis significance testing. It evaluates the consistency of the data with multiple competing hypotheses (Burnham and Anderson, 2002). In the current example, there are only two possible hypotheses: the frog is absent (H_a) and the frog is present (H_p). Likelihood-based methods ask: 'What is the probability of observing the data under each of the competing hypotheses?' In this example it is the probability of not detecting the species during a visit to a site.

Unlike null hypothesis testing, likelihood-based methods, including information-theoretic methods, do not consider the possibility of more extreme (unobserved) data. The likelihood for a given hypothesis can be calculated as the probability of obtaining the data given that the hypothesis is true.[4] Despite the implication of its name, the likelihood of a hypothesis is not the same as the probability that the hypothesis is true.

Under the first hypothesis (the frog is absent), the probability of observing the data ($\Pr(D \mid H_a)$) is equal to 1. Under the second hypothesis (the frog is present) the probability ($\Pr(D \mid H_p)$) is 0.2. Information-theoretic methods then determine the amount of evidence in favour of these two hypotheses by examining the ratio of these values (Burnham and Anderson, 2002).[5] These ratios may be interpreted by rules of thumb (see also Chapter 4). Using the criteria of Burnham and Anderson (2002),

---

[4] The likelihood need only be proportional to the probability of obtaining the data, not strictly equal to it. Terms that do not include the data or the parameters being estimated can be ignored because they will cancel out of the subsequent calculations.

[5] Information-theoretic methods are modified by the number of parameters that are estimated with the data. In this case, the parameter of the analyses (the detection rate) is not estimated with the data, so the number of estimated parameters is zero.

we might conclude that the southern brown tree frog is 'considerably less' likely to be present than it is to be absent $(\Pr(D \mid H_p)/\Pr(D \mid H_a) = 1/5)$.

Bayesian methods

Frequentist methods are in general not well-suited to the species detection problem because they are strictly limited to assessing long-run averages rather than predicting individual observations (Quinn and Keough, 2002). This is revealing; frequentist methods are not strictly suitable for predicting whether a species is absent from a particular site when it has not been seen. Such a problem is fundamental in ecology, which relies on knowing the distribution of species. In contrast, the species detection problem can be tackled using Bayesian methods.

Bayesian methods are similar to likelihood-based methods, but also incorporate prior information using what is known as 'prior probabilities'. Bayesian methods update estimates of the evidence in favour of the different hypotheses by combining the prior probabilities and the probabilities of obtaining the data under each of the hypotheses. The probability that a hypothesis is true increases if the data support it more than the competing hypotheses.

Why might the prior information be useful? If the researcher visited a pond that appeared to have excellent habitat for southern brown tree frogs (e.g. a large well-vegetated pond in a large leafy garden), then a failure to detect the species on a single visit would not necessarily make the researcher believe that the frog was absent. However, if the researcher visited a pond that was very unlikely to contain the frog (e.g. a concrete fountain in the middle of an asphalt car park), a single failure to detect the frog might be enough to convince the researcher that the southern brown tree frog did not occur at the pond. Frequentist methods cannot incorporate such prior information, but it is integral to Bayesian methods.

Another key difference between Bayesian methods and frequentist methods is that instead of asking: 'What is the probability of observing the data given that the various hypotheses are true?' Bayesian methods ask:

What is the probability of the hypotheses being true given the observed data?

At face value, this is a better approach for our problem because we are interested in the truth of the hypotheses (the frog's presence or absence at the site) rather than the probability of obtaining the observed data given different possible truths.

Example 1 9

In practice, Bayesian methods differ from likelihood methods by weighting the likelihood values by the prior probabilities to obtain posterior probabilities. I will use the two symbols $Pr(H_a)$ and $Pr(H_p)$ to represent the prior probabilities. Therefore, the likelihood for the presence of the frog given that it was not seen (0.2) is weighted by $Pr(H_p)$ and the likelihood for the absence of the frog (1.0) is weighted by $Pr(H_a)$. Thus, the posterior probability of presence is a function of the prior probability $Pr(H_p)$, the data (the frog was not seen) and the model, which describes how the data were generated conditional on the presence or absence of the frog. Now we must determine a coherent scheme for determining the values for the prior probabilities $Pr(H_p)$ and $Pr(H_a)$. This incorporation of prior information is one of the unique aspects of Bayesian statistics. It also generates the most controversy.

Both hypotheses might be equally likely (prior to observing the data) if half the sites in the study area were occupied by southern brown tree frogs (Parris unpublished data). In this case, $Pr(H_a) = 0.5$, as does $Pr(H_p)$. With these priors, the probability of the southern brown tree frog being absent will be proportional to $0.5 \times 1.0 = 0.5$, and the probability of it being present will be proportional to $0.5 \times 0.2 = 0.1$.

The posterior probabilities must sum to one, so these proportional values (0.5 and 0.1) can be converted to posterior probabilities by dividing by their sum $(0.5 + 0.1 = 0.6)$. Therefore, the probability of the frog being present is $1/6$ $(= 0.1/0.6)$, and the probability of absence is $5/6$ $(= 0.5/0.6)$. So, with equal prior probabilities $(Pr(H_a) = Pr(H_p) = 0.5)$, we would conclude that the presence of the frog is five times less probable than the absence of the frog because the ratio $(Pr(H_p \mid D)/Pr(H_a \mid D))$ equals $1/5$. You may have noticed that this result is numerically identical to the likelihood-based result. I will return to this point later.

A different prior could have been chosen for the analysis. A statistical model predicts the probability of occupancy of ponds by southern brown tree frogs based on the level of urbanization (measured by road density), characteristics of the vegetation, and the size of the pond (based on Parris 2006.). If the pond represented relatively high-quality habitat, with a predicted probability of occupancy of 0.75, then the probability of the frog being present will be proportional to $0.75 \times 0.2 = 0.15$ and the probability of absence will be proportional to $(1 - 0.75) \times 1.0 = 0.25$. With these priors, the probability of the frog being present is equal to $3/8$ $(= 0.15/(0.15 + 0.25))$, and the probability of absence is $5/8$ $(= 0.25/(0.15 + 0.25))$.

The incorporation of prior information (the presence of good quality habitat) increases the probability that the pond is occupied by southern brown tree frogs compared to when the prior information is ignored (0.375 versus 0.167). The actual occupancy has not changed at all – the pond is still either occupied or not. What has changed is the researcher's belief in whether the pond is occupied. These Bayesian analyses may be formalized using Bayes' rule, which, following a short introduction to conditional probability (Box 1.2), is given in Box 1.3.

Box 1.2
Conditional probability

Bayes' rule is based on conditional probability. Consider two events: event C and event D. We are interested in the probability of event C occurring given event D has occurred. I will write this probability using the symbol $\Pr(C \mid D)$, and introduce three more symbols:

$\Pr(C)$ – the probability of event C occurring;

$\Pr(D)$ – the probability of event D occurring; and

$\Pr(C \text{ and } D)$ – the probability of both events occurring together.

Conditional probability theory tells us that:

$$\Pr(C \text{ and } D) = Pr(D) \times \Pr(C \mid D),$$

which in words is: the probability of events C and D both occurring is equal to the probability of event C occurring given that event D has occurred multiplied by the probability of event D occurring (independent of event C). The $\mid$ symbol means 'given the truth or occurrence of'.

The above can be rearranged to give:

$$\Pr(C \mid D) = \Pr(C \text{ and } D) / \Pr(D).$$

For example, *Pfiesteria*, a toxic alga is present in samples with probability 0.03 (Stow and Borsuk 2003). *Pfiesteria* is a subset of *Pfiesteria*-like organisms (PLOs), the latter being present in samples with probability 0.35. Therefore, we can calculate the conditional probability that *Pfiesteria* is present given that PLOs are present:

$$\Pr(\textit{Pfiesteria} \mid \text{PLO}) = \Pr(\textit{Pfiesteria} \text{ and PLO}) / \Pr(\text{PLO})$$
$$= 0.03/0.35 = 0.086.$$

Example 1 11

Box 1.3
Bayes' rule for a finite number of hypotheses

Conditional probability (Box 1.2) states that for two events C and D:

$$\Pr(C \text{ and } D) = \Pr(D) \times \Pr(C \mid D).$$

C and D are simply labels for events (outcomes) that can be swapped arbitrarily, so the following is also true:

$$\Pr(D \text{ and } C) = \Pr(C) \times \Pr(D \mid C).$$

These two equivalent expressions for $\Pr(C \text{ and } D)$ can be set equal to each other:

$$\Pr(D) \times \Pr(C \mid D) = \Pr(C) \times \Pr(D \mid C).$$

It is then straightforward to obtain:

$$\Pr(C \mid D) = \Pr(C) \times \Pr(D \mid C) / \Pr(D).$$

Let us assume that event C is that a particular hypothesis is true, and event D is the occurrence of the data. Then, the posterior probability that the frog is absent given the data ($\Pr(H_a \mid D)$) is:

$$\Pr(H_a \mid D) = \Pr(H_a) \times \Pr(D \mid H_a) / \Pr(D).$$

The various components of the equation are the prior probability that the frog is absent ($\Pr(H_a)$), the probability of obtaining the data given that it is absent ($\Pr(D \mid H_a)$, which is the likelihood), and the probability of obtaining the data independent of the hypothesis being considered ($\Pr(D)$).

The probability of obtaining the data (the frog was not detected) given H_a is true (the frog is absent) was provided when using the likelihood-based methods:

$$\Pr(D \mid H_a) = 1.0.$$

Similarly, given the presence of the frog:

$$\Pr(D \mid H_p) = 0.2.$$

The value of $\Pr(D)$ is the same regardless of the hypothesis being considered (H_p the frog is present, or H_a the frog is absent), so it simply acts as a scaling constant. Therefore, $\Pr(H_a \mid D)$ is proportional to $\Pr(H_a) \times \Pr(D \mid H_a)$, and $\Pr(H_p \mid D)$ is proportional to

$\Pr(H_p) \times \Pr(D \mid H_p)$, with both expressions having the same constant of proportionality $(1/\Pr(D))$.

$\Pr(D)$ is calculated as the sum of the values $\Pr(H) \times \Pr(D \mid H)$ under all hypotheses. When prior probabilities are equal $(\Pr(H_a) = \Pr(H_p) = 0.5)$:

$$\Pr(D) = [\Pr(H_a) \times \Pr(D \mid H_a)] + [\Pr(H_p) \times \Pr(D \mid H_p)]$$
$$= (0.5 \times 1) + (0.5 \times 0.2) = 0.6.$$

Therefore, the posterior probabilities are $5/6$ $(0.5/0.6)$ for the absence of the frog, and $1/6$ $(0.1/0.6)$ for the presence of the frog.

So, for a finite number of hypotheses, Bayes' rule states that the probability of the hypothesis given the data is calculated using the prior probabilities of the different hypotheses $(\Pr(H_j))$ and the probability of obtaining the data given the hypotheses $(\Pr(D \mid H_j))$:

$$\Pr(H_i \mid D) = \frac{\Pr(H_i) \times \Pr(D \mid H_i)}{\sum_j \Pr(H_j) \times \Pr(D \mid H_j)}$$

This expression uses the mathematical notation for summation $\sum$.

If on the other hand, the pond had poor habitat for southern brown tree frogs, the prior probability of presence might be 0.1. Thus, $\Pr(H_p) = 0.1$ and $\Pr(H_a) = 0.9$. As before, $\Pr(D \mid H_p) = 0.2$ and $\Pr(D \mid H_a) = 1.0$. Note that the values for the priors but not the likelihoods have changed. Using Bayes' rule (Box 1.3), the posterior probability of presence is:

$$\Pr(H_p \mid D) = \Pr(H_p) \times \Pr(D \mid H_p) / [\Pr(H_p) \times \Pr(D \mid H_p) + \Pr(H_a) \times \Pr(D \mid H_a)]$$
$$= 0.1 \times 0.2 / [0.1 \times 0.2 + 0.9 \times 1.0]$$
$$= 0.022$$

Therefore, there is only a small chance that the frog is at the site if it has poor habitat and the species is not detected on a single visit.

Bayesian methods use probability distributions to describe uncertainty in the parameters being estimated (see Appendix B for more background on probability distributions). Probability distributions are used for both priors and posteriors. The frog surveying problem has

Example 1 13

two possible outcomes; the frog is either present or absent. Such a binary outcome (e.g. presence/absence, heads/tails, increasing/decreasing) can be represented by a Bernoulli probability distribution, which is a special case of the binomial distribution with a sample size of one. Bernoulli random variables take a value of one (representing the presence of the frog) with a probability equal to p and a value of zero (representing the absence of the frog) with probability $1-p$. Therefore, uncertainty about the presence of the frog at the pond can be represented as a Bernoulli random variable in which the probability of presence is equal to p.

It is important to note that a probability distribution is used to represent the *uncertainty* about the presence of the frog. The frog is assumed to be actually present or absent at the site, and the distribution is used to represent the probability that it is present. There appears to be misunderstanding among at least some ecologists that Bayesian parameters do not have fixed values, but change randomly from one measurement to another. Although such models can be accommodated within Bayesian analyses (e.g. by using hierarchical models, Box 3.6), parameters are usually assumed to have fixed values. The prior and posterior distributions are used to represent the uncertainty about the estimate of the parameters.

I have illustrated three components of a Bayesian analysis: priors, data and posteriors. I have not explicitly stated the model, which is the fourth aspect I mentioned in the introduction. The model in the above example is relatively simple and is the same as was used in the frequentist analyses. It can be stated as: 'the detection of the southern brown tree frog during the survey occurs randomly with a probability (p_{detect}) that depends on whether the pond is occupied ($p_{\text{detect}} = 0.8$) or not ($p_{\text{detect}} = 0.0$)'.

This model may be written algebraically as:

$$p_{\text{detect}} = 0.8 \times \textit{present}$$
$$\textit{detected} \sim \text{Bernoulli}(p_{\text{detect}}).$$

The second expression says that the variable called '*detected*' is a Bernoulli random variable. A value of one for '*detected*' indicates that the frog was detected and a zero indicates it was not. The probability of detection is equal to p_{detect}, and is given in the first equation. It depends on whether the frog is present at the site (*present* $= 1$, $p_{\text{detect}} = 0.8$) or absent (*present* $= 0$, $p_{\text{detect}} = 0.0$).

Random sampling from the posterior distribution using WinBUGS

This Bayesian analysis can also be implemented in the freely available software package WinBUGS (Spiegelhalter *et al.*, 2005). Appendix A provides information about obtaining the program WinBUGS and a tutorial on its use. I will use WinBUGS throughout the book, so it is worth investing some time in understanding it. Readers who are unfamiliar with WinBUGS should study Appendix A now, before continuing with the rest of the book.

The acronym WinBUGS is based on the original program BUGS (Bayesian inference Using Gibbs Sampling), but is now designed to run under the Microsoft Windows operating system (hence the Win prefix). WinBUGS works by randomly sampling the parameters used in Bayesian models from their appropriate posterior distributions. Because the posterior distribution for the example of detecting southern brown treefrogs can be calculated (Box 1.3), it is not necessary to use WinBUGS in this case. However, for many problems it is difficult or impossible to calculate the posterior distribution, but samples from it can be obtained relatively easily using WinBUGS or other MCMC software. If a sufficiently large number of replicates are taken, the form of the posterior distribution can be determined and its parameters, such as the mean, standard deviation, and percentiles, can be estimated.

WinBUGS takes samples from the posterior distribution by using 'Markov chain Monte Carlo' (MCMC) methods. 'Monte Carlo' implies random sampling, referring to roulette wheels and other games of chance. 'Markov chain' refers to the method of generating the random samples. A series of random numbers in which the value of each is conditional on the previous number is known as a Markov chain. MCMC algorithms are constructed in such a way that the samples from the Markov chain are equivalent to samples from the required posterior distribution (see Appendix C).

The advantage of using Markov chains for sampling from the posterior distribution is that it is not necessary to calculate the value of the denominator in Bayes' rule. The calculation is avoided because each successive sample depends on the ratio of two posterior probabilities that share the same denominator, which then cancels (Appendix C). This simplifies matters, because the Bayesian analysis only requires the product of the prior probability and the likelihood of the data.

Example 1 15

If each sample depends on the value of the previous sample, successive values drawn from the Markov chain may be correlated. Correlations between the samples have some important consequences. The first is that the initial values that are used in the Markov chain may influence the results until a sufficiently large number of samples is generated. After this time, the 'memory' of the initial values is sufficiently small and the samples will be drawn from the posterior distribution (Box 1.4). Because of the potential for dependence on the initial values, their possible influence is

---

Box 1.4
The burn-in when sampling from Markov chains

It can take thousands of iterations for some Markov chains to converge to the posterior distribution, while others converge immediately. Therefore, it is necessary to check convergence, and discard the initial samples from a Markov chain until convergence is achieved. These discarded values are referred to as a 'burn-in'.

There are several ways to check for convergence. One of the simplest is to plot the sampled values versus the iteration number. In the example in Fig 1.3, the initial value is approximately 1200, changing to values in the approximate range 100 to 400 after five samples. The values continue to be around 100 to 400 indefinitely, suggesting that the chain has reached what is known as its stationary distribution. The Markov chain is constructed in such a way for Bayesian analyses that this stationary distribution is the posterior distribution (Appendix C).

A further check for stationarity is to initiate the Markov chain with a second set of initial values. The stationary distribution will be insensitive to the initial values. If two chains with different initial values converge, then it suggests that both chains have reached their stationary distribution. There are formal methods for checking the convergence of a pair of Markov chains, such as the Gelman-Rubin statistic (Brooks and Gelman, 1998), which compares the variation of the samples within chains and the variation of the samples when the chains are pooled. Initially, the pooled variation will be greater than the average variation of the two chains and then become equal as the chains converge. Additionally, the level of variation both within and between chains should stabilize with convergence.

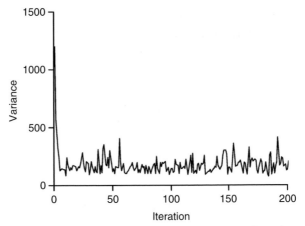

Fig. 1.3 The first 200 samples of the variance of the number of trees in a remnant for the model in Box 3.2.

examined and it may be necessary to discard some of the initial samples (perhaps the first few thousand or more) as a 'burn in' (Box 1.4).

A second consequence of any correlation is that, compared to an uncorrelated sample, each additional sample contains only a fraction of the information about the posterior distribution. Because of this, a large number of samples may be required to obtain a sufficiently precise sample if there is strong correlation between samples. Although the presence of correlation in the Markov chain reduces the efficiency of the sampling algorithm, it does not preclude the use of Markov chain methods. The reduced efficiency is simply the cost to be paid when it is not possible to obtain an analytical solution for the posterior distribution. Gilks *et al.* (1996) provides further information about Markov chain Monte Carlo methods.

The frog surveying problem in WinBUGS

Code for analysing the frog surveying problem in WinBUGS is given in Box 1.5. A Bayesian model specified in WinBUGS has the four components of a Bayesian analysis:

- prior distributions for the parameters being estimated;
- data;
- a model that relates the parameters to the data; and
- the posterior distributions for the parameters.

Example 1 17

Box 1.5
WinBUGS code for determining the presence of a species

The frog surveying problem involves determining whether the species is present at a site given that it was not detected during a survey. In WinBUGS, the code works by specifying the prior for the probability of presence and the model, which describes how the parameter of interest (the presence of the frog) is related to the data. Pseudo-code for this problem would be:

1. Specify the prior probability of presence;
2. Specify that the frog is either present or absent with a particular probability;
3. Calculate the probability of detecting the species given that it is either present (probability of detection $= 0.8$) or absent (probability of detection $= 0.0$);
4. Cpecify that the detection of the frog or failure to detect the frog (the data) arises randomly, depending on the probability of detection.

Steps 1–2 specify the prior for the presence of the frog. Steps 3–4 specify the model, describing how the data (the observation of an absence in this case) are related to the presence of the frog, which is the parameter being estimated.

The WinBUGS code for the frog surveying problem is written below.

```
model
{
  prior <- 0.5                      # the prior
                                      probability of
                                      presence
  present ~ dbern(prior)            # actual presence
                                      drawn from a
                                      Bernoulli
                                      dist'n
  prob_detect <- 0.8*present        # prob of
                                      detection depends
                                      on
                                      presence/absence
```

```
detected ~ dbern(prob_detect) # actual detection
                                   occurs with
                                   random variation
}
list(detected = 0)                 # the data - the frog
                                      was not detected
```

In this model we are interested in determining whether the frog is present (represented by the variable present). The variable prior is the prior probability of the frog being present. The prior probability of the frog being absent is therefore 1−prior. The actual presence at the site is determined randomly, by drawing from a Bernoulli distribution; a value of one indicates the frog is present and zero indicates the frog is absent. Therefore, the first two lines define the expected presence of the frog prior to the collection of the data.

The next two lines describe the model of how the data were collected. If present, the probability of detecting the frog (prob_ detect) is equal to 0.8, and it will equal zero if it is absent. The fourth line then states that the data are assumed to occur randomly, again drawn from a Bernoulli distribution, with the probability of detecting the frog on a single visit being equal to prob_detect, and the probability of not detecting the frog being equal to 1−prob_detect.

The observed data (written in the line list(detected = 0)) then influence the values of the variable present, through the application of Bayes' rule within WinBUGS (Box 1.3). Values of the variable present are sampled by WinBUGS such that they are drawn as random samples from its posterior distribution. Sampling in this way is called Monte Carlo sampling. It is a relatively common method of analysing probabilistic models (Box 1.6). If enough samples are taken, the probability of the frog being present can be estimated by the proportion of times that the variable present equals one. This proportion equals the mean of the variable present.

Sampling 100 000 times from this model in WinBUGS (after ignoring the first 10 000 samples) leads to a mean value of present of 0.17, which is equivalent to 1/6, as determined analytically. This is our estimate of the posterior probability that the site is occupied given the prior and the data. Changing the value of prior to 0.75 leads to a mean value of present that is equal to 0.38 (again based on 100 000 samples), which is equivalent to 3/8 as determined analytically.

Example 1 19

The results in WinBUGS are not exact because of random sampling error. If we took more samples in WinBUGS, the results would be closer to the truth. For example, the posterior probability of presence equals 0.3754 if half a million samples are taken when the prior for this value is 0.75. It is not precisely the same as the true answer (0.3750), but the answer in WinBUGS will continue to become more precise as more samples are taken.

Box 1.6
Monte Carlo methods

Monte Carlo methods use simulation to estimate the probability of occurrence of uncertain events. For example, consider a five-card poker hand. We could use probability theory to work out the chance of obtaining a flush (five cards of the same suit). The probability is equal to the probability that the second, third, fourth and fifth cards are the same suit as the first. For a 52-card deck, this is equal to:

$$(12/51) \times (11/50) \times (10/49) \times (9/48) = 0.00033$$

We could also work out this probability with a Monte Carlo method by dealing, shuffling, and re-dealing and calculating the proportion of times that a flush appears. If we did this ten times, we might get one flush (if we were lucky). Based on these results (one occurrence out of ten deals), we might estimate that the probability of a flush is 0.1. This is an imprecise estimate. Obtaining more samples increases the precision. If we dealt the cards 10 000 times, we might get three flushes, implying that the probability of a flush is 0.0003. This is better, but still not perfect; we could deal the cards several million times and get an even more precise estimate of the probability.

Of course, it is laborious to deal the cards that many times. An efficient alternative might be for a machine to deal the cards for us. Such a task might be suitable for computers, because they specialize in repetitive tasks. However, instead of dealing a physical deck of cards, the computer could use its circuitry to generate 'random' numbers that have the same statistical properties as the cards. In this case, thousands of samples can be generated very quickly by randomly generating integers between 1 and 52 (representing the 52 possible cards) with equal probability.

This virtual random sampling is the same sort of process that is used by WinBUGS. It generates samples that have the same statistical properties as the posterior distribution. The samples generated by WinBUGS can then be analysed to estimate the statistical properties of the posterior distribution such as its mean and percentiles.

WinBUGS code includes the prior for the parameters, but most of the code is usually the model, which describes how the data are related to the parameters. The posterior is then generated by WinBUGS with Monte Carlo sampling (Box 1.6; Appendix C).

The advantage of using a Monte Carlo approach is that it is able to sample from the posterior distribution without analysts having to do the various calculations themselves. In the frog surveying problem, the calculations done by hand are relatively easy. In the few cases where the calculations can be done by hand, they are usually more difficult, and in most other cases they are impossible.

Monte Carlo methods have another appealing property. Even relatively complex statistical analyses (e.g. regression analysis) do not require WinBUGS code that is much more complex than that presented in Box 1.5. Once familiar with relatively simple analyses, it is not much more difficult to write code for more complex analyses.

Example 2: Estimation of a mean

The second example of Bayesian analysis involves estimating the average diameter of trees in a remnant patch of eucalypt forest (Harper *et al.*, 2005). The size of trees is important when studying, for example, nutrient dynamics, provision of habitat for animals, production of nectar, mitigation of temperature extremes, and amelioration of pollution (Bormann and Likens, 1979; Attiwill and Leeper, 1987; Huang, 1987; McPherson *et al.*, 1998; Brack, 2002; Gibbons and Lindenmayer, 2002; Brereton *et al.*, 2004).

The mean diameter of trees could conceivably take any value between zero and some large number. Therefore, the hypotheses are not discrete. There are an infinite number of hypotheses, represented by any conceivable value for the mean diameter of the trees. Bayesian methods are able to accommodate these sorts of cases where hypotheses are distributed

Example 2 21

along a continuum, by using continuous rather than discrete probability distributions to represent uncertainty in the variables. The only modification to Bayes' rule is how the constant of proportionality is calculated (Box 1.7).

Assume that a researcher has measured the diameter of 10 randomly selected trees. In analysing the data, the researcher must choose the prior probability distribution for the parameters being estimated. Although the mean size of trees could be conceivably any positive number, the researcher has previously measured more than 2500 trees in 43 other

---

Box 1.7
Bayes' rule for continuous hypotheses

In the case of continuous hypotheses, continuous probability distributions are used to represent different possible values for parameters. Bayes' rule is then expressed as:

$$\Pr(H \,|\, D) = \frac{\Pr(H) \times \Pr(D \,|\, H)}{\int_0^\infty \Pr(x) \times \Pr(D \,|\, x)dx},$$

where H represents a particular value for the parameter. The integral in the denominator substitutes for the summation in the discrete case, and the limits of the integration are over all the possible values of the parameter (x), which in this case is assumed to be positive. This integral makes Bayesian methods difficult to conduct analytically, because in most cases it cannot be determined.

Readers who are uncomfortable with mathematics may look at the above equation and decide that they can never solve those sorts of problems and decide that Bayesian methods are too hard. The complexity of the equation should not be discouraging because in most cases it is impossible to solve, regardless of a person's mathematical skills. Fortunately, software is available so users do not need to evaluate or even construct the integral.

As with the case when there were a finite number of hypotheses (Box 1.3), the denominator simply acts as a scaling constant, because it is the same for all possible values for the parameter H. As with discrete hypotheses, the posterior probability is simply proportional to the prior probability $(\Pr(H))$ multiplied by the likelihood $(\Pr(D \,|\, H))$. The main analytical task of Bayesian analyses is to determine the constant of proportionality.

remnants (Harper *et al.*, 2005). After measuring so many trees in the study area, he has a good idea about likely values for the mean diameter of trees in the previously unmeasured remnant. Frequentist analyses do not permit this additional information to be used in determining the mean diameter of trees in the new remnant, but a Bayesian analysis does.

Based on data from the other 43 remnants, the mean diameter of trees in remnants is 53 cm and the mean varies among remnants with a standard deviation of approximately 5 cm. Assuming the mean diameter of trees follows a normal distribution, we would expect approximately 95% of remnants to have a mean tree diameter that is within 1.96 standard deviations of the overall average. Therefore, prior to collecting the data there is a 95% chance that the mean diameter of trees in the new remnant will be between approximately 43 and 63 cm. This prior reflects the researchers' expectation of the mean size of trees in a newly measured remnant based on his previous experience in the study area. A plot of the prior shows the range of likely values (Fig. 1.4).

The Bayesian solution for the normal mean

In the simplest case, and to make the analysis comparable to a traditional frequentist analysis, we will assume that the diameter of trees within the

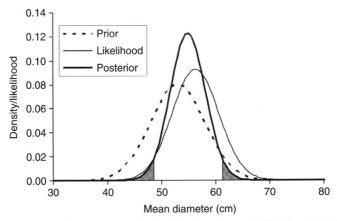

Fig. 1.4 The prior and posterior density functions and likelihood for the mean diameter of trees in a remnant, based on a sample of ten trees. The posterior would equal the likelihood if the prior was uninformative. The posterior is more precise than both the prior and the likelihood function because the posterior combines the information in both. The limits of the 95% credible interval of the posterior have 2.5% of the area under the curve in each tail (shaded).

Example 2 23

remnant follows a normal distribution. In the case where the data and the prior both have normal distributions, Bayes' rule (Box 1.7) provides an analytical solution for the posterior distribution. However, analytical solutions are available for only a handful of Bayesian models, so I will first illustrate this example using WinBUGS (Box 1.8). It is simply a matter of specifying a prior distribution for the mean of the diameter

---

Box 1.8
Estimating a mean for a normal model using WinBUGS

In estimating the mean diameter of trees, the prior has a mean of 53 cm and a standard deviation of 5 cm. In WinBUGS, the width of a normal distribution is expressed using the precision ($1/\text{variance} = 1/\text{sd}^2$), which in this case is equal to 0.04 ($1/25 = 1/5^2$).

In this example, the variance of the data is assumed to be known, making it equivalent to using a z-value rather than a t-value in a frequentist analysis. However, uncertainty in estimating the precision of the data can be included easily in the WinBUGS analysis (Chapter 3).

The pseudo-code for the WinBUGS analysis is:

1. Specify the prior for the mean diameter of trees in the remnant as being normally distributed with a mean of 53 and precision of 0.04 (standard deviation of 5).
2. Calculate the standard deviation of the data.
3. Specify the precision of the data as the inverse of the variance of the diameter of trees in the remnant (the variance equals 184.9 in this example).
4. For each of the ten trees that were measured, assume that their diameter is drawn from a normal distribution with the mean and precision as specified in steps 1 and 3.

The WinBUGS code is:

```
model
{
   m ~ dnorm(53, 0.04)        # prior for mean
   stdev <- sd(Y[])           # calculate std deviation
                                of data
```

```
prec <- 1/(stdev*stdev)  # precision of the data =
                           1/variance
for (i in 1:10)           # for each of the ten
                           trees ...
{
   Y[i] ~ dnorm(m, prec)  # diameter drawn from a
                           normal distribution
}
}
list(Y = c(42, 43, 58, 70, 47, 51, 85, 63, 58, 46))
```

The 'for loop', designated by the line for (i in 1:10) and subsequent line within the curly brackets, is equivalent to ten lines of code, one for each of the ten trees, i.e. Y[1] ~ dnorm(m, prec), up to Y[10] ~ dnorm(m, prec). It is shorthand to replace repetitive sections of code.

The data are provided in the line:

```
list(Y = c(42, 43, 58, 70, 47, 51, 85, 63, 58, 46))
```

The 'c' before the brackets indicates that the following data are concatenated (linked together) into the one variable, with the first variable represented by Y[1], the second by Y[2], etc.

This analysis also requires that the user specifies an initial value of mean for the Markov chain (Box 1.4). The choice is not important because the chain converges quickly to the posterior distribution in this case, and could be generated randomly. However, in some cases the speed of convergence is increased if the Markov chain is initiated with values that are close to the posterior distribution, so the following arbitrary value was used:

```
list(m = 55)       # an arbitrary initial value
```

After discarding the first 1000 samples as a burn-in, 100 000 samples were generated in WinBUGS. Of these samples, 2.5% were less than 48.5 and 2.5% were more than 61.3. Therefore the 95% credible interval is 48.5−61.3 cm for the mean diameter of trees in the remnant. This result is insensitive to the choice of the initial value.

Example 2 25

of trees in the remnant, and then constructing a model in which the measured diameters are drawn from a distribution with that mean. The posterior distribution calculated in WinBUGS is the same as that obtained using the analytical solution (Box 1.9).

Confidence intervals and credible intervals

A frequentist analysis would ignore the prior information and simply use the mean of the data and the standard error $(=\sqrt{(184.9/10)}=4.3)$,

---

Box 1.9
Estimating a mean for a normal model analytically

When the data and prior have normal distributions, the posterior distribution also has a normal distribution, the mean and variance of which depends, not surprisingly, on the mean and variance of the prior. The posterior distribution also depends on the sample size, mean and variance of the data. The mean and variance of the posterior can be calculated from the following formulae (Gelman *et al.* 2004):

$$\mu_{post} = \frac{\mu_{prior}/\sigma^2_{prior} + \mu_{data}n/\sigma^2_{data}}{1/\sigma^2_{prior} + n/\sigma^2_{data}}, \text{ and}$$

$$\sigma^2_{post} = \frac{\sigma^2_{prior}\sigma^2_{data}/n}{\sigma^2_{data}/n + \sigma^2_{prior}},$$

where n is equal to the sample size, σ^2_{prior}, σ^2_{data}, and σ^2_{prior} are the variances of the prior, data and posterior, and μ_{prior}, μ_{data} and μ_{post} are the means of the distributions.

These formulae provide useful insights into Bayesian statistics. The mean of the posterior is a weighted average of the means of the prior and data. The weights are the precisions of the prior $(1/\sigma^2_{prior})$ and the data (n/σ^2_{data}). The influence of the data and prior on the posterior mean depends on which is more informative. When there are no data $(n=0)$, the mean of the posterior is equal to the mean of the prior. When the variance of the prior is very large, $1/\sigma^2_{prior}$ approaches zero and the mean of the posterior will be close to the mean of the data. The prior is said to be uninformative when the posterior is influenced exclusively by the data. This is achieved by using a prior with a large variance.

The variance of the posterior has similar properties, but these are most obvious when its formula is re-arranged to be expressed as the inverse of the variance:

$$\frac{1}{\sigma^2_{post}} = \frac{n}{\sigma^2_{data}} + \frac{1}{\sigma^2_{prior}}$$

The inverse of the variance measures precision. Large values for the precision mean the variance is small. The quantity n/σ^2_{data} is the inverse of the standard error squared, and it measures precision in an ordinary frequentist analysis. Therefore, the precision of the posterior is simply equal to the precision based on the data (the inverse of the standard error squared) plus the precision of the prior. The precision of an estimate is increased by using prior information.

The diameter measurements of ten trees in the new remnant (42, 43, 58, 70, 47, 51, 85, 63, 58, 46 cm) have a mean of 56.3 and variance of 184.9. Given the prior has a mean and variance of 53 and 25, the posterior distribution for the mean diameter of trees in the new remnant has the following mean and variance:

$$\mu_{post} = \frac{53/25 + 56.3 \times 10/184.9}{1/25 + 10/184.9} = 54.9$$

$$\sigma^2_{post} = \frac{25 \times 184.9/10}{184.9/10 + 25} = 10.6$$

The standard deviation of the posterior is 3.26 cm ($\sqrt{10.6}$). Therefore, there is an approximate 95% chance that the mean diameter of trees in the park is between 48.5 cm and 61.3 cm (the mean of the posterior plus or minus 1.96 times the standard deviation of the posterior) (Fig. 1.4). This 95% credible interval is the same as that obtained from WinBUGS (Box 1.8).

---

leading to a 95% confidence interval of 47.9−64.7 cm (56.3 $\pm$ 1.96 $\times$ 4.3). This is the same as the credible interval that was obtained when using a Bayesian analysis with an uninformative prior (Box 1.10).

Bayesian credible intervals and frequentist confidence are usually numerically identical if the Bayesian prior is uninformative. An

Example 2 27

Box 1.10
Estimating the mean of a normal model with an
uninformative prior

An uninformative prior for the mean diameter of trees can be
specified by using the following line of code for the prior instead of
the one in Box 1.8:

```
mean ~ dnorm(0, 1.0E-6)    # wide prior for mean
```

This is a very wide normal distribution with a mean of zero and a
standard deviation of 1000. Therefore, mean diameters between, for
example, zero and 200 cm have approximately the same prior
probability. If this uninformative prior is used, the posterior
distribution for the mean diameter of trees in the remnant has a mean
of 56.3 and 95% credible interval of 47.8−64.7, numerically
equivalent to the 95% confidence interval of a frequentist analysis.

---

uninformative prior is one in which the data (via the likelihood, which is
$Pr(D \mid H)$ in Bayes' rule) dominates the posterior. This is achieved
by using a prior with a large variance. A large variance permits the
parameter to be drawn from a wide range of possible values and the prior
probabilities of all reasonable parameter values are approximately equal.
When the prior distribution is uninformative, the posterior distribution
has the same form as the likelihood (Fig. 1.5). The likelihood and
posterior have different forms when the prior is informative (Fig. 1.4).

The posterior distribution is less precise, and hence the credible
interval is wider, if the prior information is ignored (Fig. 1.5). Ignoring
the prior information would imply that the researcher believed that
the remnant could have any mean diameter prior to collecting the data.
Such a belief would be inconsistent with the researchers' previous
experience in the study area, which provides useful data on the range of
likely results.

Although frequentist confidence intervals and Bayesian credible
intervals may appear similar, they are in fact different. For a 95%
Bayesian credible interval, there is a 95% chance that the true value of the
parameter will be within the interval. Ecologists are often interested in
this kind of interval because they want to know the chance that the true
value of the parameter is within a specified range. Such an answer
requires the use of Bayesian credible intervals.

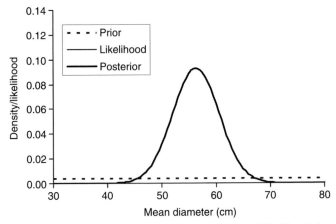

Fig. 1.5 The prior and posterior density functions and likelihood for the mean diameter of trees in a remnant, based on a sample of ten trees and using an uninformative prior. The prior distribution (drawn on an arbitrary scale to assist comparison) has a mean of zero and standard deviation of 1000 making mean diameters between 30 and 80 cm all equally likely a priori. The likelihood and the posterior are indistinguishable. The posterior is less precise than in Fig. 1.4 because the prior is uninformative.

In contrast, a 95% frequentist confidence interval does not contain the true parameter with 95% probability. Instead, it is based on the concept of an infinite number of samples. If I repeat the data collection an infinitely large number of times and construct 95% confidence intervals for the mean for each set of data, 95% of these confidence intervals would encompass the true mean.

This different meaning of confidence and credible intervals is not just semantic. In some circumstances, it can lead to numerical differences even when the credible interval is based on an uninformative prior. For example, in estimating a fail-safe period from three observations of failure times (12, 14 and 16), Jaynes (1976) shows that the shortest possible 90% confidence interval is 12.1–13.8. This interval does not contain the true fail-safe period, which must be less than the smallest observed lifespan (12). This result is not an error. The method of calculating 90% confidence intervals will produce intervals that enclose the true value of the parameter 90% of time. However, the true value might surely lie outside any single interval, as in this example.

In contrast, the Bayesian analysis with an uninformative prior arrives at a sensible conclusion; the shortest possible 90% credible interval is 11.2–12.0 (Jaynes, 1976). When the intervals are the same, the choice of

Bayesian or frequentist methods does not matter. However, when the intervals are different, only Bayesian methods provide logical results (Jaynes, 1976).

Concluding remarks

In introducing Bayesian methods, this chapter made two important points. Firstly, Bayesian methods can answer questions that are relevant to ecologists, such as: 'What is the probability that this hypothesis is true?' and 'What is the probability that a parameter will take values within a specified interval?' Secondly, relevant prior information can also be incorporated into Bayesian analyses to improve the precision of estimates.

Bayes' rule is the basis of Bayesian methods. It is derived as a simple expression of conditional probability. The rule specifies how prior information and data are combined using a model to arrive at the posterior state of knowledge. Both the prior and posterior states of knowledge are represented as probability distributions. The posterior probability simply equals the prior probability multiplied by the likelihood of the data and a scaling constant. Bayesian methods become difficult because the scaling constant is usually hard to calculate analytically. However, recent numerical methods such as Markov chain Monte Carlo make Bayesian methods accessible to all scientists.

Frequentist confidence intervals and Bayesian credible intervals will usually be numerically equivalent if uninformative priors are used. In this way Bayesian methods provide a numerical generalization of frequentist methods. They also do so in such a way that probabilistic statements about the state of nature are mathematically logical. The next chapter provides a more thorough comparison of different statistical schools and examines their various strengths and weaknesses.

2

Critiques of statistical methods

Introduction

Statistics in the discipline of ecology is dominated by null hypothesis significance testing. Apart from the construction of confidence
intervals, it is almost the only statistical method taught in ecology
at the undergraduate level. In leading ecological and conservation
journals, such as *Conservation Biology*, *Biological Conservation*, *Ecology*
and the *Journal of Wildlife Management*, null hypothesis testing has
been used in approximately 90% of articles between 1978 and 2001
(Anderson *et al.*, 2000; Fidler *et al.*, 2004), although this proportion was only 80% in 2005 (Fidler, 2005). Since 1980, there have
been several thousand null hypothesis tests (on average) reported
each year in *Ecology* (Anderson *et al.*, 2000), a further illustration of
the dominance of this method. In comparison, only about 5% of
ecological articles refer to Bayesian methods and even fewer use them
(Fig. 1.1).

Despite its dominance, null hypothesis significance testing has
ardent critics. There are alternatives but their use is controversial. In this chapter, I review three different methods of statistical
analysis that are used in ecology (see also Oakes, 1986). These are
null hypothesis significance testing, information-theoretic methods,
and Bayesian methods. Readers will not be surprised, given the
topic of this book, that I believe there are clear advantages in
using Bayesian methods, although not necessarily to the total
exclusion of others. However, I will first present an example that further
illustrates some of the differences and similarities of the statistical
methods.

Sex ratio of koalas

The following example illustrates how results of Bayesian and likelihood-based methods can differ from those obtained using null hypothesis significance testing.[1] Consider a researcher who is studying the population ecology of koalas (*Phascolarctos cinereus*) with a particular interest in the sex ratio of pouch young of mothers in poor physical condition. Assume that the researcher samples 12 female koalas in poor condition each with an offspring in its pouch (pouch young). Three of the offspring are male and nine are female. Based on this study, what can we say about the sex ratio of koalas produced by females in poor physical condition?

Null hypothesis significance testing

A reasonable null hypothesis in this case might be that the number of male and female offspring would be equal. Under this hypothesis, the sex ratio (the proportion of males in the population of pouch young) would be 0.5. However, Trivers and Willard (1973) suggest that a female-biased sex ratio would be expected in animals with poor physical condition. Thus, a reasonable alternative hypothesis is that the sex ratio is less than 0.5.

The data could have been obtained in at least two ways. Firstly, the researcher could have decided to sample 12 koalas with offspring, in which case her data are the number of males (three males). Alternatively, she may have sampled koalas until three males had been obtained, in which case her data would be the number of female koalas until the third male was encountered (nine females).[2] Regardless of the sampling strategy, the data are equivalent (three males and nine females); the only difference is the stopping rule for her sampling strategy (sample until 12 individuals, or sample until three males are obtained).

The null hypothesis tests under these two stopping rules are described in Box 2.1. Under the first stopping rule the *p*-value is 0.073, so using the

---

[1] This example is based on a thought experiment conducted by Lindley and Phillips (1976). It has been modified from tossing of coins for an ecological audience (Johnson, 1999).

[2] This second sampling strategy may seem odd at first, but it may occur in reality when a researcher has multiple questions. For example, the study might be mainly based on researching male offspring, with the calculation of the sex ratio as a secondary interest. Additionally, the sample sizes may be regarded by some as very small (three males or 12 offspring), but I chose them for illustration because this simplifies the mathematics.

Box 2.1
Null hypothesis tests for a proportion

In the first case, 12 offspring are sampled so that the number of males can take any number between zero and 12. If the sex ratio (the proportion of the population of pouch young that are male) were r, the chance of getting 0 males would equal the probability of the first being male (r), multiplied by the probability of the second being male (r), etc. Thus, the probability of all 12 koalas being male equals r^{12}.

The probability of one koala being male and the other 11 being female is equal to the probability that the first is male and all the others are female $r(1-r)^{11}$, plus the probability that the second is male and all others are female $r(1-r)^{11}$, etc. Thus, the probability that only one of the 12 is male is equal to $12r(1-r)^{11}$.

It turns out that the sampling in this case can be described by the binomial distribution (Appendix B, Johnson *et al.*, 1992, Fowler *et al.*, 1998), which states that the probability of there being x males in a sample of 12 is given by:

$$\text{Pr(males} = x) = \frac{12!}{(12-x)!x!}r^x(1-r)^{12-x},$$

where $x!$ ('x factorial') equals $1 \times 2 \times 3 \times \ldots \times x$, with $0! = 1$.

The p-value for the null hypothesis $r = 0.5$ is the probability of getting three males or a more extreme result (in this case, fewer than three males) from a sample of 12. Thus:

$$P_1 = \text{Pr(males} = 3) + \text{Pr(males} = 2) + \text{Pr(males} = 1) + \text{Pr(males} = 0).$$

After substituting the binomial probabilities, one obtains $P_1 = 0.073$.

Thus, using the conventional 'cut-off' (type I error rate) of 0.05, we would conclude that the sex ratio *is not* significantly (in a frequentist sense) less than 0.5.

What if the researcher used the different sampling strategy in which she records the sex of koala pouch young until three males have been sampled? The probability of there being no females in the sample is equal to the probability that the first three pouch young are male (r^3).

One female will be sampled if one of the first three pouch young is female while the fourth is male at which point sampling will cease. Thus, the probability that there is one female in the sample is equal to the probability that two of the first three pouch young are male (and one is female) ($= (3!/1! \times 2!)r^2(1-r)$ from the binomial distribution) multiplied by the probability that the fourth is also male (r), leading to $3r^3(1-r)$.

The probability that there are two females in the sample is equal to the probability that two of the first four pouch young are male (and two are female) ($= (4!/2! \times 2!)r^2(1-r)^2$) multiplied by probability that the fifth is male (r), leading to $6r^3(1-r)^2$. This can be continued indefinitely for any number of females.

More generally, the number of females until the ith (in this case third) male is described by the negative binomial distribution (Appendix B; Johnson *et al.*, 1992; Fowler *et al.*, 1998). Its probabilities are given by:

$$\Pr(\text{females} = x) = \frac{(3 + x - 1)!}{x!2!}r^3(1 - r)^x.$$

Using this sampling design, the *p*-value is equal to the probability of sampling nine or more females before the third male, and it is given by:

$$P_2 = \Pr(\text{females} = 9) + \Pr(\text{females} = 10) + \Pr(\text{females} = 11) + \cdots$$

The series continues indefinitely (until 'females' equals infinity), because there is the (small) possibility that three males will not be sampled even after sampling many animals. Substituting the negative binomial probabilities into the above equation leads to $P_2 = 0.033$.

If we again used the conventional type I error rate of 0.05, we would conclude that the sex ratio is significantly (in a frequentist sense) less than 0.5.

usual type-I error rate of 0.05 we would not reject the null hypothesis that the sex ratio is equal to 0.5. The *p*-value is 0.033 under the second stopping rule, so we would accept the alternative hypothesis that the sex ratio is less than 0.5.

The two different stopping rules for the sampling strategies lead to different conclusions about the null hypothesis, even though the actual data are identical. Therefore, our decision about the sex ratio of koala pouch young would be determined not only by the data we collected but by how we decided to stop sampling. The difference occurs because the null hypothesis test depends on the data *and more extreme* (but unobserved) values. The stopping rule does not influence the results if the data are analysed using Bayesian or information-theoretic methods because their results are not conditioned on unobserved data (Lindley and Phillips, 1976; Berger and Berry, 1988). I will illustrate the information-theoretic and Bayesian solutions below.

Information-theoretic methods

Information-theoretic methods use maximum likelihood estimation to determine parameter values. Maximum likelihood methods can esti-mate the sex ratio and place confidence intervals around the estimate (Edwards, 1992; Hilborn and Mangel, 1997). Maximum likelihood methods are so named because the best estimate is the one for which the probability of obtaining the observed data is maximized. Under the binomial model, the probability of obtaining the observed data (three males) is:

$$\Pr(\text{males} = x) = \frac{12!}{(12 - x)!x!}r^x(1 - r)^{12-x} = \frac{12!}{9!3!}r^3(1 - r)^9.$$

This expression is maximized when the term $L = r^3(1-r)^9$ is maximized. This occurs when the sex ratio r is equal to 0.25, for which $L_{\max} = 0.001173$.

Under the negative binomial model, the probability of obtaining the observed data (nine females) is:

$$\Pr(\text{females} = x) = \frac{(3 + x - 1)!}{x!2!}r^3(1 - r)^x = \frac{11!}{9!2!}r^3(1 - r)^9.$$

This is also maximized when the term $r^3(1-r)^9$ is maximized, illustrating that methods based on maximum likelihood are not influenced by the stopping rule.

Approximate confidence intervals can be placed on the maximum likelihood estimate of the sex ratio r by finding values of r such that L is equal to $L_{\max}\exp(-\chi/2)$, where χ is a value from the appropriate chi-squared (χ^2) distribution (Edwards, 1992; Hilborn and Mangel, 1997). For a 95% confidence interval $\chi^2 = 3.84$, which corresponds to a tail probability of 0.05 for a chi-squared distribution with one degree of freedom. Values of r for which L is equal to $L_{\max}\exp(-3.84/2) = 0.000172$ are 0.069 and 0.528. These define the limits of the 95% confidence interval for the sex ratio.

The Bayesian method

This problem can be analysed using a Bayesian method in WinBUGS (Box 2.2). Regardless of the stopping rule that is used, the probability density function for the sex ratio of offspring is the same. Therefore, only the data (and our prior) influences the estimate of the sex ratio of pouch young in koalas, not the choice of when to stop sampling. Unlike null hypothesis testing, Bayesian methods are based only on the observed data not unobserved (more extreme) values.

The mean of the posterior distribution for the sex ratio is 0.286, and the 95% credible interval is 0.091−0.537. Thus, it is likely that the sex ratio is less than 0.5, but it may in fact not be. Note that the 95% credible interval is similar to the confidence interval constructed with maximum likelihood estimation, but is different because the chi-squared value in the likelihood method requires a large-sample approximation.

This example illustrates that null hypothesis significance testing and Bayesian methods can lead to different conclusions. Additionally, when an uninformative prior is used, estimates based on Bayesian and information-theoretic methods are similar (see also Chapter 1). However, informative priors increase the precision of Bayesian estimates (e.g. Fig. 1.4). The strengths and weakness of the different statistical methods are described in more detail in the following sections.

Null hypothesis significance testing

Null hypothesis testing works in a series of steps.

1. A null hypothesis is defined, along with a single alternative hypothesis.
2. Data are collected.
3. The analyst calculates the probability of collecting the data or more extreme data given that the null hypothesis is true. This probability is the *p*-value.
4. If this probability is sufficiently small, then the analyst concludes that the data are unusual given the null hypothesis. The almost universal convention is to use an arbitrary cut-off of 0.05. If the *p*-value is less than 0.05, then the analyst concludes that the null hypothesis is unlikely to be true, and the alternative hypothesis is accepted.

Box 2.2
Bayesian analysis of a proportion

The first sampling strategy (sample 12 koalas) can be analysed from a
Bayesian perspective in WinBUGS with the following code:

```
model
{
  x ~ dbin(r, 12)  # data sampled binomially with n = 12
  r ~ dunif(0, 1) # prior for the sex ratio of pouch
                    young
}
list(x = 3)       # 3 males sampled
```

The data are given by x, and r is the sex ratio being
estimated. We assume that the data are drawn from a binomial
distribution with a sample size of 12 (see Box 2.1). For simplicity,
I have chosen a uniform distribution for the sex ratio as an
uninformative prior. This ignores the fact that a sex ratio equal
to zero or one is very unlikely to occur in any mammal species.
I could use data on the sex ratio of offspring in other
mammals or other koala populations to generate a more
reasonable prior.

Sampling 100 000 times in WinBUGS (after discarding the first
10 000 samples) provides the posterior distribution (Fig. 2.1). The
mode of the distribution is 0.25 and the median is 0.275. The mean of
the distribution is 0.286, with a 95% credible interval of 0.091−0.537.
The posterior distribution indicates that the data are not entirely
inconsistent with a sex ratio of 0.5, but it is likely that the sex ratio is
less than 0.5, in accordance with the Trivers and Willard (1973)
hypothesis.

The second sampling strategy (sample until three males have been
recorded) can also be implemented in WinBUGS. In this case the
value for the number of females before the third male is encountered
is drawn from a negative binomial distribution (see Box 2.1 and
Appendix B). The WinBUGS code is:

```
model
{
  x ~ dnegbin(r, 3) # number of females sampled neg.
                      binomially
```

```
r ~ dunif(0, 1) # prior for the sex ratio of pouch
                young
}
list(x = 9)        # 9 females sampled before the 3rd
                male
```

Again, sampling from WinBUGS provides the posterior distribution. In the negative binomial case, the result is identical to the sampling strategy that used the binomial model (Fig. 2.1). Therefore, only the data (and our prior) would influence the estimate of the sex ratio of pouch young in koalas, not the choice of how to stop sampling.

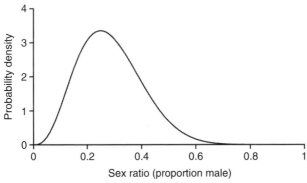

Fig. 2.1 Posterior probability density function for the sex ratio of koalas, based on a sample of three males and nine females and a uniform prior between 0 and 1.

5. If the probability is not below the critical level, then the analyst fails to reject the null hypothesis. By implication, the null hypothesis is 'accepted', but it is not proved because null hypothesis significance testing can only falsify hypotheses.

Define the null hypothesis and its alternative

The null hypothesis is a statement about the state of the system, often expressed in terms of parameter values. For example, an arbitrarily chosen null hypothesis[3] is that the Shannon-Weiner index of plant species diversity is the same in salt, brackish and fresh water marshes (Mullan Crain *et al.*, 2004). An alternative hypothesis is also chosen, which will be accepted if the null is rejected. In this example, the alternative hypothesis is that the plant species diversity is different in marshes of different salinity.

The choice of a useful null hypothesis is important. Ideally, the null hypothesis should be such that its rejection will have important logical consequences that lead to better ecological understanding (Underwood, 1997). However, ecologists routinely use nil nulls (predicting no effect or no difference) that are very unlikely to be correct (Johnson, 1995; Anderson *et al.*, 2000). These hypotheses are also referred to as false or trivial null hypotheses, or silly nulls (Stephens *et al.*, 2005). Anderson *et al.* (2000) reported that 90% of ecological studies use silly nulls.

Silly nulls take forms such as 'the survival of juveniles and adults is the same', 'there is no relationship between two variables of interest', or 'the growth rate of individuals is the same' (Anderson *et al.*, 2000). While studies that include silly nulls can provide useful scientific information (e.g. by demonstrating the size of effects), the rejection of a trivial null hypothesis is largely worthless because it was not a reasonable proposition in the first place.

The above null hypothesis of Mullan Crain *et al.* (2004) could be viewed as trivial. A priori we would expect that the diversity index for plants in marshes will vary with the salt content of water. As ecologists, we know fresh and salt water marshes would contain different plant species, and will therefore almost certainly have different diversity indices. A fundamentally more interesting question might be about how the diversity index changes across the salt gradient. Mullan Crain *et al.* (2004) do address this, but as it is not concerned with null hypothesis testing I will return to it later.

Why do ecologists use nil nulls so frequently when their rejection is usually uninformative? Why do we bother trying to reject literally thousands if not millions of hypotheses each year that are probably false? Perhaps because it is difficult to construct null hypotheses with a non-zero effect. For example, in the study of Mullan Crain *et al.* (2004), an alternative null hypothesis is difficult to formulate a priori. Although we can be reasonably sure that a difference would be expected, it is difficult to specify a precise prediction for a non-nil hypothesis. A theory

---

[3] This example was chosen as the first null hypothesis encountered after a randomly
selected page of the 2004 volume of the journal *Ecology*.

that could predict species diversity of marshes as a function of the salt content of the water would provide a reasonable null hypothesis. Rejection of the null in this case would be very interesting, because it would tell us that the theory is lacking.

Such falsification of a well-reasoned hypothesis is a potentially powerful aspect of null hypothesis testing. However, ecological theory is not sufficiently precise that exact null hypotheses (other than nil nulls) can be constructed routinely. There are some exceptions, such as allometric models that predict particular scaling exponents (West *et al.*, 1997). However, the prevalence of nil nulls suggests that similarly precise predictions for non-nil nulls are rare in ecology. Although there is a large amount of data available to ecologists, such data can at best be used to make uncertain (probabilistic) predictions. Null hypothesis testing, like other frequentist methods, is not suitable for evaluating predictions that are imprecise.

Ecologists may also use null hypothesis testing through an adherence to Popperian falsification (e.g. Underwood, 1997). However, the rejection of a trivial null hypothesis fails to meet Popperian, or any other well-known philosophical criteria for good scientific practice. Further, Popperian falsification can be achieved without null hypothesis significance testing. If null hypothesis testing is to be used successfully, ecologists need to use logical null hypotheses. The evidence demonstrates that this does not occur despite continued criticism of the use of null hypothesis testing in ecology (Johnson, 1995; Anderson *et al.*, 2000; Fidler *et al.*, 2006).

Collect data

There is little that is controversial when it comes to collecting data for null hypothesis testing. It is assumed that the subjects, quadrats or other units of sampling are selected at random, while accounting for any underlying stratification or structure in the data during the analysis. Similar or identical assumptions apply to any statistical method. Readers should refer to literature on experimental design for further information (Underwood, 1997; Quinn and Keough, 2002).

Calculate the *p*-value

The *p*-value of null hypothesis testing is equal to the probability of obtaining the observed data *or more extreme data* if the null hypothesis is true. For example, consider the null hypothesis that the exponent of the scaling relationship between metabolic rate and body size is 0.75. Then, we collect some data on metabolic rate and body size and estimate the value of the exponent as 0.77, leading to a difference between the null hypothesis and the estimate of 0.02. However, given the variation expected in the data, a difference this large might be expected just by chance. The *p*-value is the probability of getting a difference this big or bigger if the null hypothesis is true.

Critics of null hypothesis testing ask: 'Why should data that have never been observed (e.g. the occurrence of an exponent greater than 0.77) influence our inference about the validity of the null hypothesis?' This seems to be a reasonable concern. It is easy to construct examples in which the observed data are impossible if the null hypothesis is true, but where the *p*-value is not zero because more extreme data are possible (e.g. a null hypothesis of an odd number of breeding birds in a monogamous species).

In practice, most null hypotheses predict unimodal distributions for the data, with the most common form being a normal distribution or a similar distribution derived from the normal (e.g. t or chi-squared distribution). As a result, there is usually a monotonic relationship between the probability of obtaining the observed data and the *p*-value. As the probability of observing the data increases, so too does the *p*-value. Therefore, the influence of the 'unobserved results' is usually small. For the example in Box 2.1, the different stopping rule led to different interpretations of what constituted more extreme data. The subsequent difference in the *p*-value was relatively small (0.033 versus 0.073), although large enough that the result of the hypothesis test was affected in this case.

Reject the null hypothesis if the *p*-value is small

A small *p*-value indicates that the observed data would be unlikely to occur if the null hypothesis were true. This then provides evidence against the null hypothesis, and it will be rejected and the alternative hypothesis accepted if the *p*-value is sufficiently small. The logic of this process is not entirely straightforward and is often misinterpreted. The most common misinterpretation is that the *p*-value is the probability of the null hypothesis being true, given the data. This misinterpretation is even shared by some of those who teach the method (Haller and Krauss, 2002). However, it is actually the converse of this; it is the probability of data (or more extreme data) given that the null hypothesis is true (Berger and Sellke, 1987; Ellison, 1996).

The distinction between the two probabilities can be illustrated with an example of probability with which readers will be familiar. Consider the null hypothesis that I am rolling a fair six-sided die. Let the observed data be that a value of one is rolled. The p-value for this outcome is 0.167 (1/6). Now consider the inverse of this problem: if a one is rolled, what is the probability that I am using a fair six-sided die? It is definitely not 0.167. The probability that I am using a fair die would depend on whether I own and use biased dice. Your belief in whether I am using a fair die has more to do with the perception of my character than the result of a single throw of the die.

However, if I continued to roll ones on subsequent throws, you would be rightly suspicious. The important point is that the probability of obtaining the data and the probability of a hypothesis being true are not the same, although there is a relationship between the two. This relationship is defined by Bayes' rule (Box 1.3).

Null hypotheses are routinely rejected when the p-value is less than an arbitrary value of 0.05. This choice has virtually no basis in logic. It is simply a number that ensures that correct null hypotheses would be rejected only 5% of the time in the long run. This rate of rejection of correct null hypotheses is the type I error rate. The type II error rate is the proportion of times that false null hypotheses would not be rejected. Power is equal to one minus the type II error rate. It is the proportion of times that false null hypotheses are rejected.

Ideally, the probability of making a poor decision with null hypothesis testing should decline to zero as the sample size increases. In fact, it would be possible to ensure this is the case if statistical power was considered by ensuring that both the type I and type II error rates were reduced towards zero as the sample size increased. However, by having a slavish adherence to the threshold of 0.05, ecologists set a limit such that even the largest studies will lead to erroneous conclusions about true null hypotheses 5% of the time.

When the null hypothesis is rejected, we accept the alternative hypothesis without explicitly considering how well it matches the data. Unless the alternative hypothesis is constructed with care and is a reasonable choice, we run the risk of accepting a hypothesis that is even more implausible (given the data) than the null we just rejected. Box 2.3 provides an example where the apparent rejection of a hypothesis and acceptance of an unlikely alternative has caused considerable trouble. The same data are analysed using information theoretic methods and Bayesian methods. Only Bayesian methods arrive at the correct conclusion in this case.

Box 2.3
Null hypotheses in the courts

Two sons of Sally Clark, a London lawyer, died while very young about a year apart and both in mysterious circumstances. In 1998, seven months after the second death, Sally Clark was charged with murder. She was eventually tried, found guilty, and sentenced to two life terms of imprisonment in 1999.

Part of the evidence presented in her trial was that the probability of two children dying of cot death in the one family was vanishingly small (quoted in court as one in 73 million). This is essentially a p-value: the probability of the obtaining the data (two children dying) given the null hypothesis (Sally Clark was innocent). Since this probability is so small, the null hypothesis of innocence could be rejected and the alternative hypothesis (that Sally Clark murdered her children) accepted. As claimed by the prosecutor, two cot deaths were 'beyond coincidence'. Of course, this acceptance of the alternative hypothesis ignores whether the available evidence supports it and whether it is reasonable in the first place.

The application of null hypothesis testing in this case gets it alarmingly wrong. Despite the vanishingly small p-value, evidence came to light that demonstrated that Sally Clark was unlikely to have killed her two sons, and after spending more than three years in prison she was released. A small p-value does not necessarily mean that the alternative hypothesis is true.

Fail to reject the null when the p-value is large

If the p-value is large, it would be nice to be able to conclude that the null hypothesis is true. Large p-values can occur if the null hypothesis is true or close enough to being true. However, they can also occur if the study is not sufficiently well-designed to have a reasonable chance of generating a low p-value if an important difference from the null hypothesis actually exists. Even a p-value of 1.0, which is the highest value it can possibly be, does not necessarily provide strong evidence that the null is true because large p-values can also be obtained if the null is false but the study is poorly designed.

The quality of a study is measured by its statistical power, and p-values need to be interpreted in its light. Power is the probability of obtaining a statistically significant result given that the null hypothesis is not true. Statistical power can help determine necessary sample sizes and assist the planning of data collection and subsequent analysis. The only problem is that this is rarely done in ecology. Power is almost never reported by ecologists, but in approximately half of all cases authors interpret their non-significant results as evidence that the null hypothesis is true (Peterman, 1990; Taylor and Gerrodette, 1993; Johnson, 1999; Anderson *et al.*, 2000; Fidler *et al.*, 2004). This is despite the fact that power must be known if we are to interpret the importance of non-significant results (Fidler *et al.*, 2004).

When calculating power it is necessary to specify both the difference one wishes to detect and the variance of the data. Both values can be difficult to determine, but any calculation of power is conditional on the values that are used. Smaller differences may go undetected, and power will be less than expected if the variance is underestimated.

Summary of null hypothesis testing

There are several problems with the use of null hypothesis significance testing in ecology. These problems are mainly due to how the method is implemented, rather than the basis of the method. In summary, errors in the use of null hypothesis testing include:

1. using silly null hypotheses;
2. believing that the p-value is the probability that the null hypothesis is true;
3. interpreting large p-values as evidence that the null hypothesis is true (a sub-set of point 2);
4. ignoring statistical power (related to point 3);
5. following the almost universal convention to use a type I error rate of 0.05, despite power being ignored, so the type-II error rate is unknown; and
6. ignoring the size of effects being estimated and/or the evidence in favour of competing hypotheses when p-values are cited in results.

There are two problems with the actual basis of null hypothesis testing:

1. Data that were never observed or cannot be obtained influence the results (e.g. Box 2.1) because the p-value is based on data that

are more extreme than those observed, as well as the observed data; and

2. Evidence in support of the alternative hypothesis is ignored in the decision about whether to reject the null hypothesis in favour of the alternative.

In practice, these two problems need not have dire consequences for null hypothesis testing. If the null and alternative hypotheses are both reasonable (i.e. there has been logical and thoughtful development of the hypotheses), then the p-value provides a measure of the evidence in support of the two possible hypotheses, although it tends to overstate the evidence against the null (Berger and Sellke, 1987; see also Chapter 4). Despite warnings, silly nulls are common in ecology, and ecologists routinely ignore power while interpreting non-significant results as evidence that the null hypothesis is true. The same errors and efforts to correct them are repeated in other disciplines (Fidler *et al.*, 2004). The evidence suggests that null hypothesis testing is used poorly. Because of these repeated problems, there has been ongoing and ardent criticism of null hypothesis testing (Parkhurst, 1997; see also http://www.warnercnr.colostate.edu/~anderson/null.html):

> Clark (1963) '... no longer a sound or fruitful basis for statistical investigation'
> Bakan (1966) '... essential mindlessness in the conduct of research.'
> Deming (1975) '... small wonder that students have trouble understanding hypothesis tests. They may be trying to think.'
> Carver (1978) '... significance testing should be eliminated; it is not only useless, it is also harmful...'
> Cohen (1994) '... hypothesis testing does not tell us what we want to know... out of desperation, we nevertheless believe that it does.'
> Rozeboom (1997) 'Null hypothesis significance testing is surely the most bone-headedly misguided procedure ever institutionalised in the rote training of scientists.'

I recommend that ecologists largely stop using it in favour of the methods discussed in the remainder of this chapter. One of these methods is Bayesian statistics. Given the problems with null hypothesis testing and its prevalence in ecology, one might ask how the discipline has managed to progress (Dennis, 1996). I believe part of the answer is that many ecologists do more than just null hypothesis testing when analysing their data. They also estimate the size of effects that they are studying. This answers more relevant questions such as 'what is the magnitude of the difference?' rather than 'is there a difference?'

This approach to data analysis is considered more fully at the end of the chapter.

Information-theoretic methods

Information-theoretic methods work in a series of steps:

1. A set of candidate models are selected that represent different hypotheses for explaining reality.
2. Data are collected.
3. The data are used to assess the relative support for the different models, by estimating the amount of information lost when using each. The best model is selected as the one that is estimated to lose the least amount of information.
4. Any required predictions are made using an average of the models that is weighted towards those that are estimated to lose less information.

Select a set of candidate models

One of the main tenets of information theoretic methods is to select a set of possible models (hypotheses) for explaining reality. Information theoretic methods are not constrained to examining only two possible hypotheses as required for null hypothesis testing. An arbitrary number of hypotheses can be examined simultaneously, but Burnham and Anderson (2002) recommend careful selection of the hypotheses. Each hypothesis is represented as a statistical model. The statistical models link the data that are to be collected to various parameter values. The models are selected with the knowledge that most, if not all models in ecology will be imperfect. The aim is to find the most parsimonious model or set of models.

An example will illustrate the construction of possible hypotheses and associated models. Grand *et al.* (1998; see also Anderson *et al.*, 2000) were interested in the effect of lead poisoning on female spectacled eider. Data were available from two sites and the birds were classified as either having been exposed to lead or not based on blood analysis. The five possible hypotheses were:

1. Survival depended on lead exposure but did not vary among sites.
2. Survival depended on both lead exposure and site, with an additive effect.

3. Survival depended on both lead exposure and site, with an interaction between the two (i.e. the effect of lead varied among the sites).
4. Survival did not depend on lead exposure but varied among the sites.
5. Survival did not depend on the site or lead exposure.

Grand *et al.* (1998) constructed statistical models for each of these hypotheses. The models related the observations of each individual over three years (the capture/recapture history) to the survival rates, and the survival rates were functions of the relevant explanatory variables. The results are presented later in this section.

Advocates of information-theoretic methods are some of the firmest critics of the (mis)use of null hypothesis testing in ecology, in particular the use of silly nulls (Johnson, 1999; Anderson *et al.*, 2000). Of course, information-theoretic methods are not immune to silly hypotheses. In the above example, it could be argued that any model that did not include an effect of lead on mortality can be discounted as unlikely a priori. Similarly, one could argue that there must be at least some difference in mortality of spectacled eiders among sites given that the birds will be exposed to different conditions (hunters, predators, food, etc.). Further, one would expect that the effect of lead poisoning on mortality would depend on the site, with no two sites having perfectly identical responses. So, we can claim a priori that model 3 is our best model and that the other models are silly. Any number of other models could be added to the list, such as those in which annual mortality varies as a function of possible weather variables.

Of course, the production of an unlimited number of increasingly complex models is counter to the aim of parsimony (finding the simplest model that still fits the data reasonably well). Advocates of null hypothesis significance testing might argue that by using silly nulls, they are doing something very similar; only including detail when there is evidence that the extra detail is justified. The difference is that information-theoretic methods are based on an aim to minimize the loss of information, whereas decisions about including parameters based on null hypothesis testing depend on the type I error rate, which is usually set at the arbitrary value 0.05.

Collect data

Again, the collection of data is a largely uncontroversial step, with principles of randomization and replication being important, as well as attempts to minimize biases and imprecision. In the spectacled eider example, the researchers marked individuals and constructed a re-sighting history that recorded whether each bird was recorded in subsequent years.

Calculate the relative amount of information lost by each model

The concept of information loss is easy to envisage for digital images. If an image is made up of many pixels, it will tend to be a good reproduction of the original scene. However, as the number of pixels decreases, the image will become less clear as the pixels become larger, and greater amounts of information (detail) will be lost. Although no digital image will provide a perfect representation of the original scene, the various images that are available can be ranked on the basis of their clarity, which measures the relative amount of information lost or gained by using one image compared to another. Similarly, some ecological models will lose more or less detail when trying to represent reality by having different levels of bias and precision.

Akaike (1973) identified the relationship between a formal measure of the information content of a model (Kullback-Leibler information) and values of the maximum likelihood or deviance that are commonly used in statistics (Anderson *et al.*, 2000). This led to Aikake's information criterion (AIC) which is an estimate of the relative Kullback-Leibler information of a model. AIC is calculated from the minimum deviance of the model (a measure of fit) and the number of estimated parameters (a measure of complexity). Poorer fitting models and more complex models lead to greater AIC values. Chapter 4 provides more information about likelihood, deviance, and AIC.

The best model, of those being considered, is the one that is expected to lose the least amount of information (i.e. has the lowest AIC value). Embedded in the calculation of AIC values is the concept of parsimony. Using AIC to select the best model involves a trade-off between model fit and complexity, with more complex models being selected only if they provide a sufficiently superior fit (see Chapter 4).

The differences in AIC among models are more important than the actual values. Differences are usually expressed relative to the model with the smallest AIC value ($\Delta_i = AIC_i - AIC_{min}$). For the spectacled eider example, the model with an effect of only lead was the best model, while the model with additive effects of site and lead was the second best model (Table 2.1)

Table 2.1. *Differences in AIC values (Δ_i) between the best model (model 1) and the other models. Model weights (w_i) were calculated using the relative AIC values (Δ_i). The effect size and standard error is based on the maximum likelihood estimate for the effect of lead for that model.*

Model	Δ_i	w_i	Effect size (s.e.)
1. Lead effect	0.00	0.673	0.337 (0.105)
2. Additive lead and site effects	2.07	0.239	0.335 (0.148)
3. Interactive lead and site effects	4.11	0.086	0.330 (0.216)
4. Site effect (no lead effect)	14.25	0.001	-
5. No site or lead effect	12.71	0.001	-

The AIC differences of each model (Δ_i) can be converted to Akaike weights (w_i) that measure the likelihood of the data given the model. When there are R candidate models, the Akaike weights are

$$w_i = \exp(-\Delta_i/2)/\sum_{r=1}^{R} \exp(-\Delta_r/2).$$

This equation simply means that the Akaike weights are obtained by transforming the AIC differences (Δ_i) using $\exp(-\Delta_i/2)$ and then re-scaling the subsequent values so that they sum to 1. Anderson *et al.* (2000) interpret these weights as approximate probabilities that the model (of those in the candidate set) is the Kullback-Leibler best model, i.e. that of the models considered it minimizes the loss of information. This means that the Akaike weights provide a measure of evidence in favour of each of the candidate models provided we have a priori reasons to believe that the models are equally reasonable. However, Box 2.4 illustrates that a priori evidence matters. Burnham and Anderson (2002, pp. 302–5) discuss more fully the relationship between Akaike weights and model probabilities.

Information theoretic methods also permit evaluation of the evidence that lead influences the mortality of spectacled eider. This is achieved by summing the Akaike weights for the models that include an effect of lead (0.998). Thus, there is very strong evidence that lead affects survival, because the models that do not include lead as an effect have very low weights.

Box 2.4
Information theoretic methods in the courts

An information theoretic approach to the evidence in Sally Clark's case arrives at a similar conclusion to null hypothesis testing (Box 2.3). In this case, we have basically two hypotheses: Sally Clark murdered her children or did not murder her children (ignoring the chance that she murdered only one of them). We can calculate the likelihood of the data (her two children died) under the two hypotheses. For the first, the probability of her two children dying given that she murdered them is clearly 1. The probability of two deaths under the second hypothesis that she is innocent can be calculated from data on cot deaths in the UK. This value is approximately one in 300 000 (not one in 73 million as quoted in court because cot deaths are unlikely to be independent events within families; Hill, 2004). We can calculate AIC values for these two hypotheses. Because the data (two sons dying) are not used to estimate the parameters of the models, $K = 0$ and the AIC values are simply equal to the deviance (-2 In(likelihood)):

Hypothesis	Likelihood	AIC	w_i
Children murdered	1.000	0	~ 1.0
Children not murdered	3.4×10^{-6}	25.2	3.4×10^{-6}

Faced with this analysis, things are not looking up for Sally Clark. The interpretation of Akaike weights (w_i) as model probabilities require us to conclude that it is likely that she murdered her two sons. However, this analysis neglects a vital piece of information, which is that parents only very rarely kill their children (Box 2.5).

Average across models

Information-theoretic methods also provide a basis for including uncertainty about the best model in assessments of effect sizes. For example, model 1 predicts that the presence of lead reduces survival by 0.337 with a standard error of 0.105 (Table 2.1). However, the other models predict slightly different effects. For example, the standard error

Box 2.5
Bayesian methods in the courts

Using the Bayesian method, the prior probability that Sally Clark murdered her two sons can be estimated from rates of infanticide in the United Kingdom (Hill, 2004). Most parents do not murder their children, so the rate of murdering two children is very low (the probability is approximately one in 2.7 million). Thus, we have the prior probability that Sally Clark murdered her two sons (0.00000037), and the probabilities of her two sons dying given the two hypotheses (1 in 300 000 if she did not murder them and 1 if she did), so we can calculate the posterior probability:

Pr(Sally Clark murdered her sons given the data)
$$= 3.7 \times 10^{-7} \times 1/[3.7 \times 10^{-7} \times 1 + (1 - 3.7 \times 10^{-7}) \times 1/300000]$$
$$= 0.1.$$

Therefore, it is approximately ten times more likely than not (given the two deaths) that Sally Clark is innocent (Hill, 2004; see also Bondi, 2004 and Joyce, 2002). Of course, other evidence could be brought to bear on this case. Firstly, the rate of infanticide among parents is much lower than the figure used if those parents do not have a history of violence towards their children, as is the case with Sally Clark. At the same time, the rate of cot death is also lower for such families (Hill, 2004). Secondly, medical evidence, some of which only came to light on appeal, increases the likelihood of death by natural causes. The first son was found to have a respiratory infection and the second a bacterial infection, both of which were likely causes of death. Thankfully, given the evidence, Sally Clark appealed her conviction and is now free after spending more than three years in jail. However, ecologists continue to fall into the same trap that appears to have contributed to the conviction of Sally Clark, which is known as the prosecutor's fallacy. This is the mistaken belief that a low probability of obtaining data given a hypothesis means that the alternative hypothesis is likely to be true.

for the effect of lead for model 3, which included the interaction term, is almost twice that of model 1 (Table 2.1).

By using model averaging, it is possible to calculate an effect of lead that accounts for uncertainty in the choice of the best model. Model averaging weights the estimated effect by the Akaike weights. Additionally, the standard error of the model-averaged predictions is a function of the within-model variation (i.e. the standard error of the prediction for each model), the between-model variation (i.e. the differences in the predictions among the different models) and the Akaike weights (see Burnham and Anderson, 2002 for details). In this example, the model-averaged prediction is that lead reduces survival by 0.335 with a standard error of 0.125.

The ability to consider more than one model when making inferences is one of the strengths of information theoretic methods. The chief advantage is recognizing that there is usually some uncertainty about which of the candidate models best describes the data. It is risky to put all one's eggs in one basket (a single model) when other plausible models might make different predictions. More detail on using multi-model inference in ecology can be found in Burnham and Anderson (2002).

Summary of information-theoretic methods

Information theoretic methods have three advantages over null hypothesis testing:

1. They are not influenced by extreme unobserved data.
2. In evaluating a hypothesis, the relative evidence in favour of the different hypotheses is assessed simultaneously while null hypothesis testing can lead to acceptance of the alternative without directly assessing evidence in its favour.
3. They permit simultaneous assessment of multiple hypotheses rather than being confined to pair-wise comparisons. Inference about the magnitude of effects can be based on the relative evidence in favour of these different hypotheses.

It has been argued that information-theoretic methods overcome the problems of null hypothesis testing. However, many of the problems of null hypothesis testing lie in its use rather than the method itself. It is entirely possible that similar errors of use may arise when using information theoretic methods (or other approaches to statistics such as

Bayesian methods). For example, the following possible errors that might arise when using this method are largely analogous to the errors that occur with the misuse of null hypothesis testing:

1. Using silly hypotheses;
2. Believing that the Akaike weight is the probability that the hypothesis is true;
3. Choosing the best model (that with the smallest AIC value) and ignoring other possible models with similar AIC values;
4. Not assessing the ability of study designs to distinguish between different models a priori;
5. Using arbitrary thresholds for differences between AIC values to decide whether a model is considered further or not;
6. Ignoring the size of effects being estimated when deciding which model is most parsimonious.

It remains to be seen whether these errors or others become common in ecology. Proponents of information-theoretic methods would argue that such errors are the fault of the user, not of the method and that the method has ways of dealing with them. A similar defence can be mounted for most of the criticisms of null hypothesis testing. Therefore, I believe that whether information theoretic methods are better than null hypothesis testing will depend on how they are used; whether people make fewer errors of interpretation and implementation when using one or the other. This is largely a question of cognition, depending on how well the different methods are taught and understood, the quality of the available software, etc.

In the next section, I describe Bayesian methods that have some clear advantages over null hypothesis testing and information theoretic methods. Bayesian methods introduce some extra difficulties, but most of these are easy to overcome.

Bayesian methods

Perhaps the main defining feature of Bayesian methods is calculation of the probability of a hypothesis being true. These hypotheses can be discrete (e.g. the frog surveying problem in Chapter 1) or continuous (e.g. when estimating a mean, Box 1.8). While both null hypothesis testing and information theoretic methods might seem to measure the reliability of different hypotheses given the data (with p-values or

Akaike weights), they actually represent the probability of obtaining the data given the hypotheses.

The steps to conducting a Bayesian analysis are:

1. A set of candidate models are selected that represent different hypotheses for explaining reality.
2. Prior probabilities are assigned to these different models.
3. Data are collected.
4. Bayes' rule is used to combine the prior probabilities with the information contained in the new data to generate the posterior predictions.

Select a set of candidate models

This is essentially the same step as used in information theoretic methods, with it being possible to use any number of competing models. The same criticisms apply. While critics point out that null hypothesis testing can lead to the use of silly nulls, there is nothing to stop silly hypotheses being used with information theoretic or Bayesian methods. Perhaps one advantage of Bayesian methods is that users are forced to establish prior probabilities for the competing models. Therefore, silly hypotheses may be noted and assigned small prior probabilities. However, how does one assign these probabilities?

Assign prior probabilities

The frog surveying problem (Chapter 1) provides an example of assigning priors to the different hypotheses. The two hypotheses are that the southern brown tree frog is present or absent from a surveyed site. If we had no previous information, then we might conclude there is nothing to choose between the two hypotheses before collecting data and assign equal prior probabilities to each. Such use of uniform priors is common in the face of ignorance.

However, if we know from previous surveys that the species is found in a particular fraction of ponds in the region, then we could use that fraction as the probability that the frog is present. The probability of the frog being absent is simply one minus the probability that it is present.

Finally, we might have a model for predicting the probability that the frog is present at ponds based on their characteristics, in which case this could be used as the prior. Each of these three cases reflects different levels of prior information. The first represents ignorance, the second the mean rate of occurrence of the frog within ponds, and the third how the rate of occurrence varies among ponds of different types. The addition of prior information in this way influences the results. If the pond has a high prior probability of the frog being present, then a single survey in which it is not seen would not be enough for us to be reasonably sure it is absent unless our ability to detect the frog was very good.

Using a uniform distribution to represent ignorance makes sense in some ways, but is problematic in others. Consider the case where we wish to determine the proportion of individuals that belong to each species in an African national park. Among the herbivores, we might be interested in the proportion of individuals that are zebras, wildebeest or some other species. By using the uniform distribution to represent ignorance, we would assign a one-third probability to each of these three classes (zebras, wildebeest, other). However, the probability of one-third is simply an artefact of our classification. The zebra would have had a probability of one-quarter if we had included gazelles as an additional class. Therefore, representing ignorance is not always straightforward.

Problems of representing ignorance also arise when specifying priors for continuous hypotheses (see also Box 3.12). For example, we may wish to estimate the density of territories of a species that are adjacent but non-overlapping. We could assign the prior distribution as uniform between 0.1 and 1.0 territories per ha if we were confident that the density was somewhere within that range but unsure of the actual value. This prior implies that the probability of the density being less than 0.2 territories per ha is 0.111 (0.1/0.9).

Alternatively, we could specify that the area of each territory is between 1 and 10 ha. This is equivalent to our limits for density (1.0 and 0.1 territories per ha, respectively). If we used a uniform distribution between 1 and 10 ha for territory size, the probability of the territory size being more than 5 ha is 0.555 (5/9), which is five times the probability calculated above for the equivalent density (0.2 territories per ha). Thus, we seem to have proved that $0.111 = 0.555$.

The difference arises because the units of the two approaches are not linearly related, so the probabilities are not conserved when they are transformed (in this case by inversion). The prior distributions are very different (Fig. 2.2). This effect of the scale of measurement is not unique to Bayesian analyses. For example, a frequentist confidence interval based on territory size would not be equivalent to a confidence interval based on densities of territories per ha.

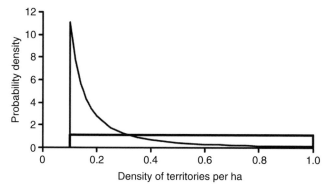

Fig. 2.2 Prior distributions for the density of territories assuming that the density is uniformly distributed between 0.1 and 1.0 territories per ha (the uniform distribution) and assuming that the size of territories (the inverse of density) is uniformly distributed between 1 and 10 ha (the sharply peaked distribution). The probability of territory density being less than 0.2 per ha (the areas under the curves to the left of 0.2) is very different for the two priors.

One of the difficulties in establishing prior probabilities is that humans tend to judge them poorly (Tversky and Kahneman, 1974; Kahneman *et al.*, 1982; Ayton and Wright, 1994; Gigerenzer and Hoffrage, 1995; Anderson, 1998; Burgman, 2005). Construction of priors by using subjective judgement is likely to depend on a range of personal attributes, how the problem is presented, motivational biases and advocacy (Anderson, 1998; Burgman, 2005). Experts are not immune to these frailties of human nature (Burgman, 2005; see also Chapter 10).

Even when there are data for constructing priors, some subjective judgement is required to determine how the prior information is represented as a probability distribution. Frequentist methods are not free of subjective judgement because they also depend on judgements about the questions to be examined, how the data are collected, the variables to be analysed, and the statistical methods and models that are used (Howson and Urbach, 1991).

It could be argued that although science is not free of subjectivity (Burgman, 2005), it should seek to minimize it (Dennis, 1996). How can Bayesian methods be used reliably and convincingly in the face of subjectivity? One approach is to be as careful, rigorous and convincing in the choice of prior as in the collection of data. This book contains various examples of using previous results and data to construct priors. Furthermore, there will be many cases where the choice of prior has virtually no effect on the results.

However, there will still be cases where uncertainty in the choice of prior remains. This uncertainty can be regarded as an honest incorporation of subjectivity in science (Berger and Berry, 1988, Howson and Urbach, 1991). The role of science is to ensure that this opinion is updated logically as evidence accumulates. Bayes' rule ensures that beliefs are updated logically, with differing opinions converging as data are collected (Cox, 1946; Howson and Urbach, 1991; Crome *et al.*, 1996; Jaynes, 2003).

There are extensions to Bayesian methods for dealing with uncertainty in the choice of priors. The methods, lumped under the title of 'robust Bayesian analysis', can also deal with uncertainty in the models used to represent the hypotheses. They involve, for example, placing bounds on the possible parameters or distributions for priors and likelihoods and, therefore, bounds on the possible posterior distributions (e.g. Berger, 1985; Walley, 1991; Ferson, 2005). Robust Bayesian methods are not without controversy, and they usually add to the computational burden. In providing an introduction to Bayesian methods for ecology, I will only touch on them briefly in Chapter 10. Interested readers are referred to Berger (1985) and Ferson (2005).

Although the prior can pose difficulties for Bayesian methods, it is in fact one of its strengths. Ecologists, in the discussion sections of journal articles, routinely consider their results in the light of previous studies. Bayesian methods provide a formal basis for these comparisons through the use of priors. Scientists may be forced to be more rigorous and less subjective when using priors to represent previous work than when simply using their judgement to make comparisons. Bayes' rule provides the means of incorporating previous findings into the formal interpretation of new data.

Use Bayes' rule to combine the prior and the data

Bayes' rule states that the probability of a hypothesis given data (the posterior) is proportional to the product of the prior and the probability of the data given the hypothesis. The constant of proportionality is given by a sum (for discrete hypotheses) or an integral (for continuous hypotheses).

There is little that is controversial about Bayes' rule itself. Given a prior probability, some data and a set of hypotheses, it provides the updated belief in the hypotheses. Bayes' rule provides a logical means (some claim the only logical means, Jaynes, 2003) of updating belief (Box 2.5).

Given that having a posterior belief in hypotheses requires a prior belief, Bayesian methods are required if we wish to use data to assign degrees of belief to hypotheses (Cox, 1946; Jeffreys, 1961; Jaynes, 2003). Bayesian methods are required even when it is difficult to construct the priors. As mentioned previously, the main controversy with Bayesian methods involves how these priors are constructed. While critics of Bayesian methods point to difficulties of establishing priors, proponents are uncomfortable about ignoring relevant prior information if it exists.

In Bayesian statistics, probability distributions for the prior and posterior distribution represent uncertainty about a parameter value. Because of this use of probability distributions, some authors refer to Bayesian parameters as random variables (Dennis, 1996; Ellison, 2004). However, this does not necessarily mean that the true value for a parameter is assumed to vary randomly from one measurement to another (Clark, 2005). The parameter might have a fixed but unknown value, which can only be expressed probabilistically. The probability distribution represents the uncertainty about the parameter, describing which values are more or less probable. As more (unbiased) data are collected, the posterior distribution becomes more concentrated on the true value for the parameter.

Arbitrarily complex statistical models can be analysed using Bayesian methods. For example, the numerical procedures for most analyses of variance require that the variance of the data for each level of a factor is identical; Bayesian analyses can easily handle cases where the variances are different. Similarly, variances can be assumed to change across the range of the data for regression analyses, rather than assuming the variance is constant. Another example is that it is relatively straightforward to introduce hierarchical effects (Clark, 2005), making it much easier to deal with problems such as pseudo-replication. Therefore, rather than making the study design conform to the required analysis, Bayesian data analyses can be made to conform to the study design. This opens the possibility of combining data from multiple sources and using a wider range of statistical models.

Nevertheless, some computational limitations of Bayesian methods remain. Computationally intensive methods (e.g. multivariate factor analyses) can take a long time to analyse with Bayesian methods, and may not be feasible if the required mathematical functions are not contained in the available software. Although these issues limit some current applications, they will most likely be surmounted with further software development.

Estimating effect sizes

One of the main types of questions asked by ecologists is how big is the effect or what is the nature of the relationship between variables? For example, we might ask how has the population size changed over time or what is the strength of the relationship between these ecological variables? Bayesian and information theoretic methods, and with slight modification the basic statistical machinery that is used to calculate p-values, can be used to answer these questions. So, rather than asking: 'Does plant diversity of marshes vary with salinity?' we could ask: 'How does plant diversity of marshes vary with salinity?' The former question was answered by Mullan Crain *et al.* (2004) using null hypothesis testing ($p < 0.0001$).

Mullan Crain *et al.* (2004) also reported the answer to the second question—species diversity increased from salt to brackish to fresh marshes. The estimated Shannon diversity indices were 0.12 ± 0.013 (mean $\pm$ s.e.) for salt, 0.22 ± 0.015 for brackish and 0.39 ± 0.016 for fresh. By providing standard errors, readers can construct confidence intervals for the estimated plant species diversity and consider the magnitude of the differences. Such considerations show that there is a trend in plant species diversity, with the diversity index in freshwater being approximately three times that of saltwater. This result is clearly more informative than 'salinity affects plant species diversity', which we expected to be true prior to the data collection and analysis. However, we might not have known the nature of those differences.

There is virtually no disagreement that the estimated size of effects and a measure of the precision of the estimate should be provided for any statistical analysis (Fidler *et al.*, 2004). The Ecological Society of America encourages this practice in their guide to authors who wish to publish in their journals. Despite this encouragement, it is not routinely practised by ecologists or by researchers in some other disciplines (Fidler *et al.*, 2004).

Frequentist confidence intervals and Bayesian credible intervals can be used to represent the precision of an estimate.[4] Such intervals can be used to determine whether an estimated effect is likely to be ecologically important. For example, McCarthy and Parris (2004) presented Bayesian 95% credible intervals for the effect of clipping toes from frogs on return

---

[4] This representation of precision is based on the length of the interval, rather than the earlier definition of precision as the inverse of the variance.

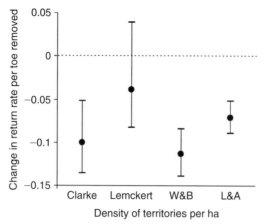

Fig. 2.3 Estimated effect of toe clipping on return rates of frogs for four different studies. The bars represent 95% Bayesian credible intervals and the circles are the means of the predicted effect. Negative values indicate an adverse effect (from McCarthy and Parris, 2004). See Chapter 8 for more details.

rates of marked animals (Fig. 2.3). The results demonstrate that toe clipping almost certainly reduces return rates in three of the four studies examined because the intervals are less than zero. Because the confidence and credible intervals in this case are numerically similar, this is equivalent to obtaining a statistically significant result, rejecting the null hypothesis that there is no effect of toe clipping (represented by the dashed line at zero).

In one study, the credible interval is rather wide and does encompass zero. This is equivalent to not rejecting the null hypothesis. However, the results can also be compared to values that might be deemed ecologically important. A value of −0.03 might be regarded as ecologically important, because given that it is not unusual to clip three toes from frogs in mark-recapture studies, the actual reduction in return rate would be ~10% per frog. If 10% of frogs are not recaptured because of the marking method, this might lead to unacceptable bias in the results as well as impacts on the population if the toe clipping is causing mortality.

By inspecting the credible intervals (Fig. 2.3), we can determine whether the results are consistent with an ecologically important effect such as −0.03. In three of the four studies, we can be confident that effects are at least this large. In the other study, there is a reasonably large chance that the reduction is greater than 0.03 per toe, but it is possible that the effect is not this large.

This illustrates one of the advantages of using intervals; the results can be compared easily to values that are ecologically meaningful, rather than only focusing on statistical significance. Although ecological importance can be examined with null hypothesis testing (e.g. by using a null hypothesis that is ecologically important), the prevalence of silly nulls means that this is rarely done.

An additional advantage of using confidence or credible intervals is that the concept of power is communicated by the width of the interval. A wide interval (relative to the size of the difference we wish to detect) means that the study has low precision—it is equivalent to having low power in null hypothesis testing. In the toe clipping example, a study would need to have a narrow confidence interval centred on a value close to (or greater than) zero to show that the adverse effect of toe clipping was not large.

A focus on ecological importance can be difficult, because we often do not know what constitutes an important difference. In these cases, one might argue that null hypothesis testing should then rely on determining the presence or absence of any difference, so a null hypothesis of no difference is appropriate. However, if we cannot establish whether an observed difference is important or not, should we be testing for any particular difference in the first place? Analyses that quantify the magnitude of effects, precision of their estimates, and the relationships among different variables would be more appropriate in these circumstances. These studies could then help us to identify ecologically important results.

In most cases, the choice between Bayesian credible intervals and frequentist confidence intervals is not important because the results are numerical similar. However, for some statistical models or when prior information is available, the two approaches will generate different answers (Jaynes, 1976; see also Chapter 1). In such circumstances, the frequentist confidence intervals perform poorly compared with the Bayesian credible intervals (Jaynes, 1976).

There are numerous calls for estimating effect sizes in ecology and other disciplines (Anderson *et al.*, 2000; Fidler *et al.*, 2004; Ziliak and McCloskey, 2004). As discussed above, there are several advantages of using intervals. Additionally, the results are much more useful for meta-analyses (Arnquist and Wooster, 1995; Gurevitch and Hedges, 2001), and the ecological importance of the results is more readily apparent. From merely selfish perspectives, this should encourage authors and editors to use intervals in their manuscripts because their work will be cited more frequently if the results are clearer and more appropriate for meta-analysis. These points relate to the progress of science, in which evidence accumulates over time. Both meta-analyses that rely on estimating effect sizes and Bayesian analyses are cumulative over studies, whereas null hypothesis testing is not.

Concluding remarks

Frequentist and Bayesian methods of statistical analysis differ in how they treat the notion of probability. Bayesian methods use probabilities to assign degrees of belief to hypotheses or parameter values. In contrast, frequentist methods (null hypothesis testing and information theoretic methods) are confined to stating the frequency with which data would be collected given hypothetical replicate sampling and specified hypotheses being true. Given the disagreement about which approach to statistics is preferred, the relative merits of the different methods are clearly a matter of opinion. My preference is for Bayesian methods because I believe ecologists are usually attempting to assign degrees of belief in parameter values, models or hypotheses more generally (Table 2.2). Ecologists regard the truth as uncertain and attempt to use science to gain an improved understanding of the truth. Such an approach is consistent with Bayesian statistics.

Many of the criticisms of the different statistical methods are directed at the use of the methods, rather than their underlying basis. Null hypothesis significance testing is criticized because of its

Table 2.2. *Benefits and limitations of Bayesian statistics (adapted from O'Hagan and Luce, 2003).*

Benefits	Limitations
Allows for intuitive interpretation	Introduces an element of subjectivity (although treating it explicitly rather than ignoring it may be a benefit)
Uses prior information	There are difficulties in constructing priors
Addresses a greater range of problems	Bayesian methods are not commonly taught
Allows complex models to be analysed easily	
Accommodates decision making	
Use all the information transparently	

widespread misuse. However, it also has logical shortfalls because unobserved data influence the results, and acceptance of the alternative hypothesis does not depend on how well the evidence supports it. Frequentist methods in general are forced to ignore any relevant prior information. Additionally, they are not well-suited to decisions about individual cases, being restricted to assessing long-run frequencies obtained from hypothetical samples. Bayesian methods are criticized because it can be difficult to determine how prior information should be incorporated into analyses.

Despite the differences, Bayesian and frequentist methods often generate numerically similar answers, especially when estimating parameters and prior information is uninformative. In these circumstances, the best approach will be largely determined by which is most easily and successfully taught, learnt, and executed. Therefore, the success of the different methods lies firmly in the realm of cognitive psychology not just statistics. However, Bayesian methods have the distinct advantage that when the numerical results differ, the Bayesian methods are invariably correct (Jaynes, 1976).

3

Analysing averages and frequencies

Ecologists routinely calculate averages and frequencies, such as the average density of a plant, or the proportion of individuals in a population that is of a particular type. In some cases, these averages are the main focus of the data analysis, while in others we are more interested in determining relationships among the items being studied and other explanatory variables. Nevertheless, analysis of averages and frequencies forms the basis of much of ecology. Therefore, I have chosen averages and frequencies as a starting point for the data analyses illustrated in this book.

The average

All ecologists are familiar with averages, particularly the arithmetic mean. The mean of a sample is calculated by summing the values observed in the sample and dividing by the sample size:

$$\bar{x} = \left(\sum x_i\right)/n.$$

To further explore the analysis of a mean, I will revisit the second example in Chapter 1, in which the following measurements of diameter (in cm) were obtained from a sample of ten trees: $42, 43, 58, 70, 47, 51, 85, 63, 58, 46$.

The average diameter of the ten trees in the sample is 56.3 cm (563/10). However, we might be more interested in knowing the mean diameter of trees in the entire remnant, which is the population from which the sample was taken. The observed average (56.3 cm) would seem like a useful estimate of it, but how reliable is it? The question of reliability of an estimate is answered by considering its precision. The usual non-Bayesian way of expressing the precision of an estimate is to calculate the standard error and perhaps also a confidence interval. In this case, the standard error is 4.3 and the 95% confidence interval is 46.6−66.0 (Box 3.1).

Box 3.1
A frequentist approach to estimating an average

A typical frequentist approach to estimating the mean is to calculate the sample mean ($\bar{x}$) and the standard error (se), determine the appropriate t-value (t) and place a confidence interval on the mean by calculating the lower ($\bar{x} - t \times se$) and upper ($\bar{x} + t \times se$) confidence limits (Fowler *et al.*, 1998). In Chapter 1 it was assumed that the standard deviation of data was known so a z-value (1.96 for a 95% interval) was used. However, uncertainty associated with the estimation of the standard deviation of the data, and hence the standard error (the standard deviation of the mean) requires the use of a t-value. This leads to a wider interval.

For the tree diameter data, the sample mean is 56.3, the standard error is 4.3, and the t-value for a 95% confidence interval is 2.262 (Fowler *et al.*, 1998). The resulting confidence interval is 46.6−66.0 cm.

A common misinterpretation of frequentist confidence intervals is that they represent a specified (in this case 95%) chance that the average diameter of trees in the remnant is between 46.6 and 66.0. This misinterpretation partly arises because researchers would often like to know the probability that the truth is within a specified interval. However, this interpretation of a confidence interval requires a Bayesian analysis.

Frequentist confidence intervals are based on the concept of multiple independent samples from the same population. If the method of sampling the population and constructing the frequentist 95% confidence interval were repeated many times, 95% of the intervals would encompass the (fixed, but unknown) true population mean. The frequentist confidence interval and the Bayesian credible interval are often numerically equivalent when an uninformative prior distribution is used. However, there is often additional information that makes the Bayesian interval narrower.

This calculation of a confidence interval is based on a model. The model states that the diameter of trees is drawn randomly from a normal distribution with a particular mean and standard deviation. Another way to think of the estimate of the mean is that it estimates one of the parameters of this model. This approach allows for a richer use of statistics—one that is based on the development of models, parameter estimation and model evaluation.

So, how might the above data be analysed using a Bayesian approach that is based on this perspective of model development, parameter estimation, and model evaluation? The first step is model development. One possible model is that the diameter of trees is drawn from a normal distribution, as used in the frequentist analysis (Box 3.1). This model has two parameters that need to be estimated, the mean and standard deviation. Using uninformative priors for the parameters, the estimated mean diameter of trees is 56.3 with a 95% credible interval of 46.6−66.0. These results are the same numerically as those based on the frequentist analyses, although they are philosophically different (Box 3.1). Given that most ecologists interpret confidence intervals as credible intervals (i.e., by assuming that they represent likely ranges for the parameter being estimated), the philosophical difference is often not important (Hoenig and Heisey, 2001, but see Jaynes, 1976).

However, confidence intervals and credible intervals differ numerically when prior information is used. In this example, prior information on the diameter of trees in 43 other remnants is available (Chapter 1). The mean diameter of trees in these other remnants has an approximate normal distribution with a mean of 53 cm and variance of 25. This distribution describes how the mean size of trees in a remnant is likely to vary among remnants. Further, prior information exists on how the variance of tree diameter varies among remnants. Some remnants have more variable tree sizes than others, but the prior information indicates the likely range of values for the variance in the remnant being measured. When the natural logarithm of the variances is taken, the transformed values have an approximate normal distribution with mean of 4.75 and variance of 0.328 (precision of 3.05). Therefore, a lognormal distribution could be used as an informative prior for the variance. The mean of the posterior would then be 55 cm with a 95% credible interval of 48.8−61.1 cm (Box 3.2).

The prediction is more precise when the prior information is included, but is it better? After measuring all the trees in the remnant (Harper *et al.*, 2005), the mean diameter turned out to be 58.7 cm, well within the

Box 3.2
A Bayesian approach to estimating an average

The normal distribution is one of the most commonly used distributions in ecology. It is often specified by two parameters, the mean of the distribution and its variance (see Appendix B on probability distributions). In WinBUGS, the two parameters used to specify the distribution are the mean and the precision, with this latter parameter being equal to 1/variance.

In specifying uninformative prior distributions for these parameters, the mean is often assumed to have an extremely flat normal distribution. In comparison, it is necessary to ensure that the precision (or variance) is positive because negative variances are not possible. Any broad continuous distribution would be suitable. This example uses a lognormal distribution.

The WinBUGS code for analysing the mean diameter of trees in the remnant is given below. It is similar to Box 1.8 but includes uncertainty in the variance of the data rather than assuming it is known. Uninformative priors are used in the code below.

```
model
{
  mean ~ dnorm(0, 1.0E-6)        # mean diameter of
                                    trees in the remnant
                                    (uninformative
                                    prior)

    var ~ dlnorm(0.0, 1.0E-6)    # variance of tree
                                    diameters
                                    (uninformative
                                    prior)

    prec <- 1/var                # converts variance
                                    to precision

    for (i in 1:10)              # for each of the ten
                                    trees
    {
       Y[i] ~ dnorm(mean, prec)  # tree diameter drawn
                                    from normal
                                    distribution

    }
}
```

```
Data
list(Y = c(42, 43, 58, 70, 47, 51, 85, 63, 58, 46))
# site 2 sample
Initial values
list(var = 100, mean = 100)
```

The for loop (`for (i in 1:10)`) means that i takes all possible values between 1 and 10, making the expression `Y[i]~dnorm(mean, prec)` insidet the loop equivalent to ten separate lines (e.g., `Y[1]~dnorm(mean, prec)` up to `Y[10]~dnorm(mean, prec)`).

Extreme values for the precision can cause numerical errors. Therefore, it is often helpful to specify initial values for the precision that are close to the posterior distribution rather than generating them randomly with WinBUGS. The first samples (1000 in this case) are discarded as a burn-in to ensure that samples from the Markov chain are not influenced by the choice of the initial values. In this case, the initial values have the approximate order of magnitude of the posterior distribution (100 and 100), but the Markov chain converges quickly to the posterior distribution.

A total of 100 000 samples from the posterior distribution provides the estimate of the mean as 56.3 cm with a 95% credible interval of 46.6−66.0, which is the same as the frequentist 95% confidence interval (Box 3.1).

However, there is prior information available for both the mean diameter of trees within remnants and the variance of the diameter. This variance of the diameter describes how different the trees are within remnants. But this variance itself varies from remnant to remnant; some remnants have more variable tree diameters than others. Based on the data from the previously visited sites, variation in the variance of the diameters can be described by a lognormal distribution, the log-transformed values of which have a mean of 4.75 and variance of 0.328 (precision = 3.05). Variation in the mean diameter among remnants can be described by a normal distribution with a mean of 53 cm and a variance of 25 (precision = 0.04). Therefore, to include this prior information, the uninformative priors are replaced by the following lines of code:

```
mean ~ dnorm(53, 0.04)    # prior with mean 53 and sd
                            5 (precision = 1/(5*5))
```

```
var ~ dlnorm(4.75, 3.05)    # informative prior for
                              variance
```

The 95% credible interval is 48.8−61.1 cm with these priors, improving the precision by reducing the width of the interval from 19.4 cm to 12.3 cm.

narrower 95% credible interval, suggesting that the prior information was useful. Although this single comparison is of limited value, it and other examples demonstrate how Bayesian methods can provide improved estimates of parameters compared to frequentist methods (O'Hagan and Luce, 2003; McCarthy and Masters, 2005).

Distributions other than the normal can be used to describe the samples. Trees cannot have negative diameters, which is possible with a normal model. However, the diameters are sufficiently large in the previous example (Box 3.2) that the normal is a reasonable approximation. The assumption of normality is evaluated in the next chapter by examining the skewness and kurtosis of the residuals.

Nevertheless, there are other cases in which the normal model is unreasonable. For example, in studying the density of trees in a New York park (McDonnell, unpublished data), the number of red oak trees in ten quadrats, each 400 m² in area, was: 6, 0, 1, 2, 1, 7, 1, 5, 2, 0.

These data are far from normally distributed because the numbers are discrete, non-negative, and the mean is not far from zero. As a first approximation, the Poisson distribution is a reasonable model because it takes only non-negative, integers (0, 1, 2, etc.), which is necessary for counts of the number of trees in quadrats (Box 3.3). It is also a reasonable choice because the Poisson distribution arises out of a model of randomness (Box 3.3). While trees are unlikely to be completely random, I will assume the Poisson distribution for now (Box 3.4). Under this model, the density of trees is estimated as 62.5 per ha with 95% credible intervals of 40−89.

This example illustrates one of the advantages of using Bayesian methods instead of statistical methods that rely on normal distributions. Here, the structure of the data influences our choice of statistical model, and it is not necessary to force the data to conform to one of the available statistical models (such as those that assume a normal distribution, equal variances, independence, etc.). This advantage is also

Box 3.3
The Poisson distribution

The Poisson distribution is appropriate for counts because the number of objects within a quadrat will follow a Poisson distribution if the objects are distributed randomly in space. The probability distribution function for the Poisson distribution is:

$$\Pr(X = x) = \frac{e^{-\lambda}\lambda^x}{x!},$$

where X is the random variable (e.g. the number of plants in the quadrat), λ is a parameter that is equal to both the mean number of plants in the quadrat and the variance in this number, and e is the constant equal to 2.71828 ... The expression $x!$ ('x factorial') is equal to $1 \times 2 \times 3 \times \ldots \times x$, with 0! equal to 1.

Thus, the probability of having no plants in a quadrat is equal to $e^{-\lambda}$ (λ^x and $x!$ are both equal to 1 when $x = 0$), the probability of one plant is $e^{-\lambda}\lambda$, the probability of two is $e^{-\lambda}\lambda^2/2$, the probability of three is $e^{-\lambda}\lambda^3/6$, etc.

Therefore, if the quadrat is 1 m^2, Fig. 3.1 shows the expected distribution of counts of the number of plants per quadrat (x) for different densities of plants (λ) ranging from 0.5–3 plants m^{-2}.

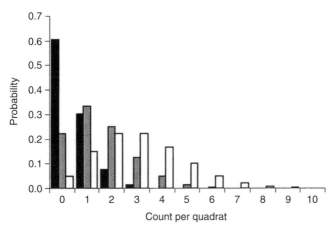

Fig. 3.1 Three Poisson distributions describing variation in the number of plants in 1-m^2 quadrats for three different plant densities: 0.5 (black), 1.5 (grey) and 3.0 (white) plants m^{-2}.

Box 3.4
Analysing the mean of the Poisson distribution

Researchers studying the structure of vegetation in New York measured the density of saplings and trees at ten quadrats (each 400 m^2 in area) in Van Cortlandt Park, one of their study sites (McDonnell, unpublished data). To estimate the average density of red oak (*Quercus rubra*) in the park, the Poisson distribution (as a model of randomness) is a reasonable description of the variation in the number of trees sampled in the quadrats.

An uninformative prior distribution for the mean density of red oak in each quadrat needs to take positive values and have a broad distribution. A lognormal distribution that had a mean of zero and standard deviation of 1000 for log-transformed values would be suitable. Using this prior, the WinBUGS code for this analysis is:

```
model
{
    for (i in 1:10)          # for each of the ten
                             quadrats
        {
            y[i] ~ dpois(m)    # number of trees drawn
                             from Poisson with mean m
        }
    m ~ dlnorm(0.0, 1.0E-6)  # uninformative prior for
                             mean trees per quadrat
}
```

Sampling 100 000 times (after an initial burn-in of 10 000 iterations is discarded) with this model in WinBUGS, using the data:

```
list(y = c(6,0,1,2,1,7,1,5,2,0))
```

and the initial value:

```
list(m = 5)
```

provides an estimate of the mean number of trees per quadrat of 2.5, and 95% credible interval of 1.6–3.6. The credible interval states that there is a 95% probability that the mean number of trees per quadrat is between 1.6 and 3.6. Because the quadrat size is 400 m^2, or 1/25th of a hectare, the density can be re-scaled to 62.5 trees per ha with a 95% credible interval of 40–89 trees per ha by multiplying the density by 25.

shared by likelihood-based methods, although it is easier to fit more complicated models with Bayesian programs such as WinBUGS (Clark, 2005).

The Poisson distribution with extra variation

The Poisson distribution was introduced in Box 3.3 as a model of randomness. Of course, plants and most other organisms may not be distributed randomly in the environment. In fact, the study of ecology is based largely on the principle that there are non-random factors that influence the distribution of organisms. Nevertheless, the Poisson distribution provides an important starting point when considering counts of organisms and other entities (e.g. Box 3.3).

One extension to the Poisson distribution is to consider that the average density of plants (equal to the parameter λ in the Poisson model) varies across the landscape. So, a randomly placed quadrat may fall into an area of high plant density or low plant density, simply as a matter of chance. We can modify the Poisson model to account for this extra source of variation by assuming that the average density of plants varies across the study area.

Any continuous distribution that is restricted to positive values might be suitable for describing the variation in the mean plant density among randomly placed quadrats. An example is analysed in Box 3.5 in which the expected number of plants per quadrat varies among quadrats varies according to a lognormal distribution. This is an example of a hierarchical model (Box 3.6).

Estimating differences

Analyses of averages can also be used to evaluate differences between paired observations. For example, Quinn and Keough (2002) discuss a study by Elgar *et al.* (1996) who analysed the effect of light on the size of orb spider webs. A total of 17 orb spiders were studied, each spinning a web in low and high light conditions. These data can be treated as a paired comparison by analysing the difference in the size of the web in low and high light conditions (Box 3.7).

Box 3.5
The Poisson model with extra variation

The data on tree density were analysed in Box 3.4 by assuming a
Poisson distribution, and are re-analysed here by assuming that the
average density of trees varies among quadrats. If it is assumed that
the variation in the average density among quadrats is described by
a lognormal distribution and sampling variation is described by a
Poisson distribution, the code can be written as:

```
model
{
  for (i in 1:10)     # for each of the 10 quadrats
  {
    mean[i] ~ dlnorm(m, tau)   # mean density drawn
                                 from lognormal
    y[i] ~ dpois(mean[i])      # no. of plants drawn
                                 from Poisson
  }
  m ~ dnorm(0, 1.0E-6)         # mean of the log
                                 density of plants
  sd ~ dunif(0, 10)            # sd of the log
                                 density of plants
  tau <- 1/(sd*sd)             # precision
}
```

A broad uniform prior is used for the standard deviation
(for the lognormal distribution in WinBUGS, this is the
standard deviation of the log-transformed density). The posterior
distribution for the standard deviation has a mean of 1.2 and 95%
credible interval of 0.33–2.9. Plausible values for the standard
deviation of log-transformed density are illustrated by considering
the complete posterior distribution (Fig. 3.2). The posterior
distribution suggests that there is likely to be variation in density
among quadrats because the standard deviation is unlikely to be
close to zero.

The median density of *Quercus rubra* can be calculated by adding
one line to the WinBUGS code (median <- exp(m)), which
back-transforms the parameter estimate for the median (which for a
normal distribution is equal to the mean) of the log-transformed
data. The median is used instead of the mean because lognormal

distributions can be highly skewed, which may make the mean much greater than the most likely values.

The 95% credible interval for the median number of *Quercus rubra* trees and saplings per quadrat is 0.43−3.9, which is wider than the 95% credible interval of the mean density when variation among quadrats is ignored (1.6−3.6, see Box 3.4). The uncertainty in the estimate has increased because the analysis permits an extra source of variation, the possibility of variation in density among quadrats. This more complicated model seems to be warranted because the posterior distribution of the standard deviation suggests its value is likely to be greater than zero (Fig. 3.2).

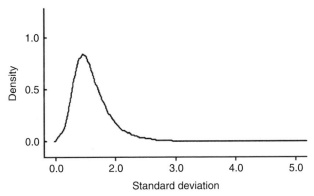

Fig. 3.2 Posterior distribution of the standard deviation among quadrats of log transformed density of *Quercus rubra* in Van Cortlandt Park, New York.

Required sample sizes when estimating means

In the previous examples, estimates of parameters were obtained by using sampled data and sometimes also by using prior information. In general, the precision of these estimates will increase as more data are collected, but how much data is necessary for a particular level of precision?

When estimating an average it is important to know what sample size is required. Ecological data can be highly variable, so sample sizes might need to be large to achieve a desired level of precision for parameter estimates. Ecological studies often suffer from insufficient sample size because unless it is calculated explicitly, scientists routinely overestimate the level of precision that they will obtain (Burgman, 2005).

Box 3.6
Hierarchical models

The mean density of plants (λ) varies from site to site for the model in Box 3.5. Rather than assuming the parameter λ is constant (as in Box 3.4), it is treated as a random variable, taking different values in each of the different quadrats. However, these values are not arbitrary numbers; the different values of λ for each quadrat are drawn from a common probability distribution, which has two parameters (m and sd).

In this example, probability distributions are used to represent two different types of uncertainty. Prior distributions are specified for the parameters m and sd, and posterior distributions are calculated (e.g. Fig. 3.2). These distributions reflect uncertainty in our estimate of parameters that have fixed values. In contrast, the mean density of plants in the quadrats is a random variable; its distribution reflects actual variation in the mean density of plants across the park.

This is an example of a hierarchical model (Fig. 3.3). The parameters of the probability distribution that describes variation among quadrats in plant density are hyper-parameters. Hierarchical models are particularly useful when adding complexity to models, for example, when considering unexplained differences in fecundity among individuals.

Rather than assuming that all individuals in a population have the same average fecundity rate, or that the fecundity rates of individuals bear no relationship to each other, we can use a hierarchical model in which the fecundity of each individual is drawn from a common probability distribution. Hierarchical models permit us to estimate the parameters of that distribution.

A practical advantage of hierarchical models is that we do not need to assume that we have described the underlying process perfectly. In the case of plant density, we might extend the model to include a prediction of how the density of oaks varies across the park, for example, as a function of soil type. But even then, it is unlikely that we will make perfect predictions of the mean. A hierarchical model would allow us to include such deterministic trends but still permit the possibility that quadrats with the same soil type would have different mean densities. This is achieved by using a parameter describing the level of variation among quadrats with the same soil type.

Hierarchical models can conceivably have an arbitrary number of levels, with hyper-parameters themselves also being treated as random quantities. In this case, the diagram in Fig. 3.3 would become a tree with an increasing amount of branching. However, the amount of data and desired level of complexity of the model will limit the number of levels that are included in the model. Further examples of hierarchical models are presented in this book, and their use in an ecological context is discussed by Clark (2005).

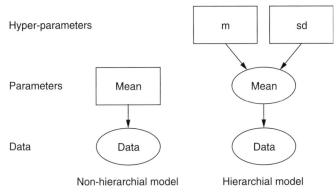

Fig. 3.3 The diagram on the left represents the generation of data under the non-hierarchical model in Box 3.4. The diagram on the right is the hierarchical model in Box 3.5. Ovals represent randomly generated variables, while rectangles represent fixed parameters that are estimated with uncertainty.

However, for non-hierarchical models, it is relatively easy to calculate the likely precision of a parameter estimate for a given level of effort.

Required sample sizes (Adcock, 1997) can be determined by calculating the precision (or variance or standard deviation) of a parameter estimate assuming that the samples are drawn from a normal distribution. In the absence of prior knowledge, the standard deviation of a mean (commonly referred to as the standard error) is equal to $s/\sqrt{n}$, where s is the standard deviation of the data and n is the sample size. If we wish to obtain a standard error of a particular magnitude (E), then the sample size must equal s^2/E^2, which is obtained by re-arranging the formula $E = s/\sqrt{n}$.

Box 3.7

Estimating the difference between paired observations

Elgar *et al.* (1996) studied the sizes of webs spun by 17 orb spiders. Each spider spun one web in high light conditions and one in low light conditions. The difference in the vertical and horizontal size of each pair of webs was determined. Using null hypothesis testing, Quinn and Keough (2002) concluded that the webs were significantly smaller in the horizontal dimension but not significantly different in the vertical dimension when spun in high light conditions.

By focusing on parameter estimation, we can measure the size of the difference. For the Bayesian analysis I assume that the differences are drawn from a normal distribution. The mean of this distribution measures the influence of light on the size of the web. Using uninformative priors that reflect a lack of prior information, the WinBUGS code for assessing the vertical difference is:

```
model
{
  vmeandiff ~ dnorm(0, 1.0E-6)    # uninformative
                                    prior for mean
                                    vert. diff.
  prec ~ dgamma(0.001, 0.001))    # uninf. prior for
                                    precision of
                                    vert. diff.
  for (i in 1:17)                 # for each of the
                                    17 spiders
  {
    VertDiff[i] ~ dnorm(vmeandiff, prec) # observed
                    diff. drawn from a normal dist'n
  }
}
```

As with all the examples, the code and data are available on the book's website. Using 100 000 samples after excluding an initial burn-in of 10 000 provides an estimate of the mean vertical difference of -20.5 cm with a 95% credible interval of -65.6–24.4. The fact that the interval overlaps zero suggests that we cannot be sure that there is no difference in the size of the webs under the different light regimes, although the estimated effect is that the webs are approximately 20 cm shorter in high light. The Bayesian credible interval is consistent with the frequentist confidence interval and *p*-value of 0.349 (Quinn and Keough, 2002).

For the horizontal dimension, the estimated reduction in web size is 46 cm, with a 95% credible interval of a 1−92 cm reduction. The credible interval, which is close to but not overlapping zero, is consistent with the frequentist *p*-value of 0.047 obtained by Quinn and Keough (2002).

This example used a gamma distribution with mean of 1 and variance of 1000 as the uninformative distribution for the precision (`prec`). The gamma distribution is commonly used as a prior for precisions because when data are normally distributed, the posterior of the precision will follow a gamma distribution when the prior has a gamma distribution. This feature simplified the computations prior to the advent of MCMC algorithms, and by convention the gamma distribution is now commonly used for precisions. As a result of a similar convention, the normal distribution is commonly used in regression (for both Bayesian and frequentist analyses).

When there is prior information in which the standard deviation of the prior is equal to *v*, the required sample size is equal to (Adcock, 1997, see also Box 3.8):

$$n^* = s^2 \left(\frac{1}{E^2} - \frac{1}{v^2} \right).$$

When the prior is uninformative, *v* is large relative to *E*, so the calculation of the required sample size approaches the value that is obtained when prior information is ignored ($1/v^2$ approaches zero). In contrast, when the prior is informative, the required sample size is reduced. For example, when the standard deviation of the prior is twice that required for the posterior, the required sample size is 25% lower than when prior information is ignored. Thus, by including prior information, it is possible to use a smaller sample size to attain the same level of precision. An example of using these formulae is provided in Box 3.9.

Propagating uncertainty in the required sample size

In calculating the required sample size in the previous examples, uncertainty in the standard deviation was ignored. If the standard deviation

---

Box 3.8
More on sample sizes

If precision is defined as the inverse of the variance, the precision of the posterior for samples drawn from a normal distribution is simply equal to the sum of the precision of the prior and the precision of the estimate based on the data alone (Chapter 1). Thus, the precision of the posterior is given by:

$$\frac{1}{E^2} = \frac{n}{s^2} + \frac{1}{v^2},$$

where s is the standard deviation of the data, n is the sample size and v is the standard deviation of the prior. The precision of the estimate based on the data alone is given by n/s^2 (the inverse of the standard error squared). Re-arranging the above equation provides the required sample size:

$$n^* = s^2 \left(\frac{1}{E^2} - \frac{1}{v^2} \right)$$

Another way of considering the problem is to think of the prior in terms of its effective sample size (m). In this case, the variance of the prior (v^2) is equal to s^2/m, and the precision of the posterior is equal to $(m+n)/s^2$. Therefore, the total sample size ($m+n$) must equal s^2/E^2, and the required (new) sample size is equal to $s^2/E^2 - m$. This is simply the sample size required in the absence of prior information minus the sample size that has effectively been collected already by using the prior.

(or its square, the variance) is underestimated, then the required sample size will also be underestimated (and vice versa). It is possible to use WinBUGS to examine how uncertainty in the variance of the data propagates through to uncertainty in the required sample size.

Uncertainty in the variance is closely linked to the chi-squared distribution when data are drawn from a normal distribution. In fact, the ratio of a sample variance to the true variance multiplied by $n-1$ has a chi-squared distribution (with $n-1$ degrees of freedom), where n is the sample size used to calculate the variance (Sokal and Rohlf, 1995). Therefore, if only the sample variance and sample size are provided,

---

Box 3.9
An example of calculating required sample sizes

Here I will illustrate the use of the formulae for the required sample size (Box 3.8). In the spider web example of Elgar *et al.* (1996, see Box 3.7), the standard deviation of the posterior for the difference in web size under high and low light was approximately 23 cm for both the horizontal and vertical dimensions. The standard deviation of the data (s) is estimated to be 94.8 cm $\left(= 23 \times \sqrt{17}\right)$. Therefore, if we wished to reduce the standard deviation of the posterior to 10 cm and used the data from the study of Elgar *et al.* (1996) as the prior, the number of new spiders required would equal:

$$94.8^2 \times (1/10^2 - 1/23^2)$$
$$= 94.8^2 \times (0.01 - 0.00189)$$
$$= 72.9.$$

Therefore, approximately 73 additional spiders would be needed to increase the precision of the estimated effect of light on the size of spider webs to the required level.

the probability distribution for the variance can be calculated. The probability distribution of the variance of the data will lead to a probability distribution of the required sample size. To ensure (with reasonable certainty) that the required precision is achieved, a greater sample size is needed to account for the possibility that the variance of the data will be greater than the value that was assumed. This feature is used in Box 3.10 to calculate the probability distribution of the required sample size for the example in Box 3.9.

The important result of the analysis in Box 3.10 is that the required sample size is underestimated if uncertainty in the standard deviation is ignored. As the sample size used to estimate the standard deviation of the data increases, the precision of the required sample size also increases and the bias decreases. However, for sample sizes that are likely to be used in pilot studies (say, $n = 10$ to 20), the bias may be important.

Adcock (1997) reviews a number of other approaches to determine sample sizes, including the use of a utility function (Lindley, 1997). Lindley's (1997) approach recognizes that extra sampling entails costs.

Box 3.10
Uncertainty in the required sample size

The sample size required to obtain a standard deviation of 10 cm for the posterior distribution of the effect of light on spider webs was estimated to be 73 (Box 3.9). However, this calculation ignored uncertainty in the standard deviation of the data, which was estimated to be 94.8 cm using a sample size of 17. In the following, it is assumed that the standard deviation is based on a sample size of 17 (hence, df = 16). Uncertainty in the true standard deviation is represented as a probability distribution by using the relationship between the chi-squared distribution and the ratio of the observed to true variance (Sokal and Rohlf, 1995).

Assuming the data are normally distributed, the WinBUGS code can be written as:

```
model
{
  req_sd <- 10              # required sd of posterior
  prior_sd <- 23           # sd of prior
  df <- 16
  x ~ dchisqr(df)          # a chi-squared variate
  r <- x/df                # ratio of observed to
                             true variance
  sigma <- 94.8/sqrt(r)    # true sd based on sample
                             sd of 94.8
  nreq <- sigma*sigma*(1/(req_sd*req_sd) - 1/
(prior_sd*prior_sd))
# number of samples required to achieve desired sd
}
```

In this case, the prior distribution has a standard deviation of 23, based on the sample size of 17. The required number of extra measurements has a wide 95% credible interval (40–168), and the mean of its posterior distribution (83) is greater than the sample size that was calculated when uncertainty in the standard deviation was ignored. If a sample size of 73 were to be chosen for additional study, there is a good chance that the required precision for the effect of light on spider webs would not be obtained. A sample size of 168 would be required to be 97.5% sure of obtaining a standard deviation of the posterior that was less than 10 cm.

Further, the benefits obtained from extra sampling depend on the objective and the amount of information already collected. Therefore, the utility function will equal the benefits minus the costs, both of which will be a function of the sample size. The optimal sample size is then chosen so that the expected value of the utility function is maximized (Lindley, 1997), or so that the probability of obtaining a minimally acceptable utility is maximized.

Estimating proportions

Ecologists are often interested in estimating proportions, such as the prevalence of disease in a population (proportion infected), the level of mortality or fecundity (proportion dying or reproducing), or sex ratios (proportion that is male). An ecologist might take a random sample of ten plants and count those that are infected with a particular disease such as *Phytophthora*. If five plants were infected, it is sensible to assume that the best estimate of the rate of infection in the population is 0.5 (5/10). What can we say about our uncertainty in this estimate?

A common way of measuring uncertainty is to calculate the standard error and a confidence interval. The most common estimate of the standard error of a proportion is (Fowler *et al.*, 1998):

$$se = \sqrt{\frac{p(1-p)}{n-1}},$$

where p is the proportion of the item in the sample and n is the sample size. The usual procedure for placing a confidence interval assumes that the estimate is normally distributed and uses a z-value to construct the confidence interval (Fowler *et al.*, 1998). For example, the 95% confidence interval for the above data would be 0.17–0.83 ($0.5 \pm 1.96 \times 0.167$). However, this is only an approximation relying on the Central Limit Theorem, so the sample size must be 'large' and the proportion not 'too far' from 0.5. For example, if only two of the plants had been diseased, the 95% confidence interval would be calculated as −0.06–0.46, which is nonsensical because a proportion can never be less than 0 or greater than 1.

It is relatively easy to analyse this problem in WinBUGS without the need for approximation (Box 3.11). If five of ten plants are diseased, the 95% credible interval is 0.23–0.77 if there is no useful prior information.

Box 3.11
Estimating a proportion

Assume we have sampled ten plants that each has a specified probability of being diseased. If the incidence of disease occurs independently among the ten plants, then the actual number of diseased plants will be a sample from a binomial distribution (Appendix B). This distribution is therefore the model used in the WinBUGS analysis. Assuming an uninformative prior for the proportion, the code would be written as:

```
model
{
  x ~ dbin(p, n)        # the number of diseased plants
                        is a binomial sample
  p ~ dunif(0.0, 1.0)   # the prior for the probability
                        of being diseased
}
```

To represent five diseased plants in a sample of ten, the data are coded as:

```
list(x = 5, n = 10)
```

Sampling 100 000 times from WinBUGS provides the 95% credible interval, which is 0.23−0.77 for five observations from a sample of ten.

This is narrower than the confidence interval that was constructed using the normal approximation. If only two diseased plants were observed in the sample of ten, the 95% credible interval would be 0.06−0.52, which is clearly superior to the interval that was based on the normal approximation. The Bayesian approach is also easier to interpret − the probability that the proportion is within the 95% credible interval is 0.95.

There are other methods for estimating proportions that do not rely on the normal approximation or Bayesian statistics. Likelihood-based methods could also be used to estimate the proportion and place confidence intervals on the estimate, resulting in a 95% confidence interval of 0.22−0.78 for the proportion 5/10, and 0.036−0.50 for the proportion 2/10. These are similar to the values for the Bayesian analysis and clearly superior to those based on the normal approximation. The difference between the likelihood approach and the Bayesian approach arises because the former uses a chi-squared approximation to construct the confidence interval. This approximation relies on the sample size being sufficiently large.

So in summary, the Bayesian approach provides logical estimates for proportions. Frequentist methods can provide useful approximations for the confidence interval, although some of these approximations may fail to provide accurate (or even reasonable) answers in some situations. Bayesian methods provide intervals with a natural interpretation (i.e. there is a 95% chance that the true parameter value lies within a 95% Bayesian credible interval), while alternative methods can only approximate this value. Why would one bother with an alternative? If you are not convinced by this simple example, consider the following use of prior information to improve a prediction.

Annual mortality of powerful owls

Powerful owls are Australia's largest owl, with the mass of females averaging approximately 1.35 kg. However, as with many owls, they are difficult creatures to study. Powerful owls occur at low population densities, are mainly active at night, and are unreliably detected using typical owl survey methods (Wintle *et al.*, 2005a). Because of this, few people have the necessary fortitude to study them. Approximately 30 powerful owl nestlings were banded in Victoria up to 1999, but none of these banded individuals were resighted as adults (McCarthy *et al.*, 1999).

McCarthy *et al.* (1999) used data from three birds that could be identified without bands to estimate annual mortality of powerful owls. One of these birds disappeared after eight years, while the other two were observed to remain alive for ten and 17 years. A single death in one year and 35 years of survival suggests the annual mortality rate is 0.03 (1/36), although given the paucity of data it is not surprising that the 95% credible interval is wide (0.0066–0.14, using $x = 1$ and $n = 36$ in Box 3.11). This interval corresponds to the owls having an average lifetime of as little as approximately seven years (1/0.14) and approximately 150 years (1/0.0066) because average lifespan is approximately equal to the inverse of the annual mortality rate. The wide confidence interval is not surprising given the relatively small amount of data.

This upper limit of 150 years appears unreasonably large. The lifespan of a single owl, let alone the average of a species, is unlikely to be as long as 150 years. Although the credible interval for mortality of

0.0066−0.14 (using a uniform prior between zero and one) is consistent with the data, it is not consistent with common sense. It is tempting to limit the survival rate to sensible values. However, it would be good to do this in a repeatable, explicit and logically consistent way. This is where Bayesian statistics can lend a hand, by using additional information to set the prior.

One approach to establishing the prior would be to use intuition to limit the possible values. Given that this is a large bird, it is unlikely that the mortality rate is greater than about 0.2, which would correspond to an average lifespan of 5 years. Similarly, it is unlikely that the average lifetime is greater than 50 years, which corresponds to an annual mortality of 0.02. Using a prior for the annual mortality that is uniform between 0.02 and 0.2 leads to a posterior distribution with a mean of 0.06 and 95% credible interval of 0.02−0.14 (Box 3.12).

In some ways, this is a perfectly good estimate. The assumptions have been clearly stated and the data have been used in a logical way to update the prior belief by using Bayes' rule. Therefore, the estimate is internally consistent with the stated logic and data. However, if the same ecologist does this calculation on a different day, he or she may arrive at a different result by deciding that 40 years would be the maximum possible average age, or 4 or 10 years as the minimum. A different ecologist is likely to use a different prior again.

The exact same ecologist using the same intuition could arrive at a different answer by using a different formulation for the prior. Instead of assuming that the mortality rate had a prior that was uniform between 0.02 and 0.2, the ecologist might assume that the average age of death was uniform between 5 and 50 years. In this case, the prior for the mortality rate would not be uninformative, but peaked near 0.02 (Box 3.12).

The posterior distribution for the mortality rate is influenced by this prior, giving a mean of 0.037 and a 95% credible interval of 0.02 to 0.09. The difference occurs because the inverse of a uniformly distributed variable will be markedly skewed (e.g. Fig. 2.2).

The potential arbitrary choice of priors is one of the limitations of Bayesian statistics. However, the potentially arbitrary influence of subjective judgement on the interpretation of results is not limited to Bayesian statistics. Faced with a conclusion that the mean lifetime of powerful owls is likely to be between 7 and 150 years, most ecologists would believe that the upper limit is unrealistically high, and would conclude that the mean lifetime is certainly less than 150 years.

Box 3.12
Estimating a proportion with a subjective uniform prior

The example of estimating the annual mortality of powerful owls using a subjectively derived prior is essentially the same as estimating a proportion (Box 3.11). However, the prior for the proportion is limited to the range 0.02 to 0.2 by using a uniform distribution. The WinBUGS code for this analysis is:

```
model
{
  x ~ dbin(m, n)          # number of deaths follows a
                            binomial distribution
  m ~ dunif(0.02, 0.2)    # a subjective prior for the
                            mortality rate
}
```

To represent that one death and 35 years of survival, the data are coded as:

```
list(x = 1, n = 36)
```

Alternatively, it could be assumed that the prior for the average age of death is uniform between 5 and 50 years. The code for this model is given below, and results in a noticeably different posterior distribution for the annual mortality.

```
model
{
  age ~ dunif(5, 50)
  m <- 1/age
  x ~ dbin(m, n)
}
```

One might wonder why a geometric model was not used to represent survival. Under the geometric model, the probability of observing the data (8 years of survival and the death) is $(1-m)^8 m$ for the owl that died. For the two owls that survived, the probabilities of observing the data are $(1-m)^{10}$ and $(1-m)^{17}$ (10 and 17 years of survival). The overall probability of observing the data is the product of these three probabilities $((1-m)^{35}m)$. This is exactly proportional to the likelihood under the binomial model with one death from 36 years, so the two approaches yield the same result.

So, without explicitly using an informative prior, subjective judgements about the reliability of the credible interval are still likely to be made. Without an explicit method for incorporating this subjective judgement, the logic will be opaque and the subjective judgement will tend to go unacknowledged.

By using Bayesian methods and a specified prior, intuition can be included in an explicit and internally consistent manner. Because the results can be sensitive to how the prior is constructed, a real challenge in Bayesian analyses is to develop informative priors that are coherent and logical. Importantly, if the prior used by an ecologist is to be viewed by others with the same confidence as the data, the method for establishing the prior needs to be clear and repeatable. The methods used to develop informative priors need to be as carefully planned, executed, and documented as the methods used to collect data. Without this detail, the rigour of the Bayesian analysis will be rightly questioned.

One possible solution to establishing a reasonable prior for the analysis of powerful owl mortality lies in the observation that mortality rates of animals vary with body mass. McCarthy *et al.* (1999) developed a regression model to predict mortality rates of powerful owls by collating data on annual adult mortality of diurnal and nocturnal birds of prey. Based on this model, the predicted annual mortality of owls is:

$$m = \frac{1}{1 + e^{0.775 + 0.954b}},$$

where b is the body mass (kg). Based on a body mass of 1.35 kg, the predicted annual mortality of female powerful owls is 0.11, with a standard error of 0.05 (McCarthy *et al.*, 1999). This prediction can be used as a coherent and logical prior, drawing on the experience of ecologists that mortality rates tend to decline in larger bodied animals.

In using the mean and standard error to establish the prior, it is necessary to choose an appropriate distribution. A normal distribution with a mean of 0.11 and standard deviation of 0.05 would not be suitable because an appreciable proportion of samples would be less than zero, which is not permitted for mortality rates. In contrast, a beta distribution with a mean of 0.11 and standard deviation of 0.05 is appropriate. Beta random variables take values between 0 and 1, making them useful priors for proportions (see Appendix B for more information on the beta distribution). The posterior distribution for the annual mortality rate, when using a beta distribution as an informative prior and confronted with the data of one death and 35 years of survival,

---

Box 3.13
Estimating a proportion with a beta prior

When using an informative beta distribution to analyse the annual mortality of powerful owls, it is necessary to determine how to calculate the required parameters. The beta distribution is defined by two parameters, *a* and *b*. Appendix B gives the formulae for calculating the values of *a* and *b* given the mean and variance of the beta distribution. A beta distribution with a mean of 0.11 and standard deviation of 0.05 requires that parameter a is equal to 4.198 and b is equal to 33.96. This then becomes the prior for the analysis of powerful owl mortality. The WinBUGS code is written as:

```
model
{
   m ~ dbeta(4.198, 33.96)      # prior with mean of
                                  0.11, and sd of 0.05
   x ~ dbin(m, n)               # assume data drawn from
                                  a binomial
                                  distribution
}
```

The corresponding data for the powerful owl analysis are again:

```
list(x = 1, n = 36).
```

After 100 000 samples from WinBUGS, the 95% credible interval for the annual mortality rate is 0.024–0.14.

---

has a mean of 0.07 (Box 3.13). The 95% credible interval for the mortality rate is 0.024–0.14.

The mortality rates of the 95% credible interval correspond to an average life expectancy of between 7 and 41 years. By using Bayesian methods, we have ensured that this estimate is logically consistent with the observed mortality rates of other birds of prey based on their body mass and the small amount of available data on powerful owls. Importantly, it provides a much more precise and meaningful estimate than using the data in isolation.

This analysis illustrates how the prior and the data combine to provide the posterior distribution for a parameter estimate (Fig. 3.4).

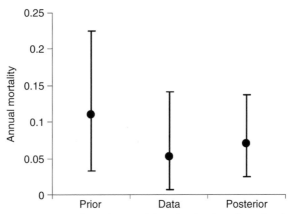

Fig. 3.4 Annual mortality of powerful owls showing the prior based on mortality estimates in other birds of prey, the data on powerful owls, and the posterior that combines the prior and the data (circles are means and bars are 95% intervals).

The posterior is more precise than the data in isolation or the prior, and the mean of the posterior (0.07) is a weighted average of the mean of the prior (0.11) and data (0.053). This averaging of the estimate and narrowing of the credible interval by including prior information is a feature of Bayesian analysis.

It is possible to use more sophisticated models both to determine the prior distribution of the mortality rate and to analyse the available data. For example, we could account for differences among species, differences among studies of the same species, and error within each study. Similarly, most data on mortality rates are derived from mark-recapture studies in which animals might not be recaptured even though they are still alive. This kind of analysis is illustrated in Chapter 7, using the example of the European dipper.

Multinomial models

The data on powerful owls allow for two possibilities for each bird (dead or alive). In such circumstances a binomial model is often appropriate. A greater number of possibilities may arise in other ecological examples. For example, habitats might be categorized into a number of different vegetation types, organisms might be classified into a number of different species, or individuals of a species might be placed into a number of age classes. The powerful owl example used a binomial model with two possible states (alive or dead). When there are more than two possible states, data can be analysed with a multinomial model.

An example of the use of the multinomial model is an analysis of the age structure of a koala population. The age structure of a population describes the distribution of individuals among age classes. This measure is important in population ecology because it can influence the rate of growth of species and the likely response to management.

The age of koalas is determined by assessing the degree of tooth wear, specifically, the degree of wear on the premolar. McLean (2003) recognized nine different tooth wear classes that define age. If we sampled a population of koalas and determined the tooth wear classes (e.g. see data in Table 3.1), what could we infer about the age structure of the population from which the sample was taken?

By using a multinomial model, in which it is assumed that each individual is assigned to one of the nine possible age classes and that each individual is a random and independent sample from the population, it is possible to estimate the age structure of the population (Box 3.14). Thus, animals in TWC II are predicted to make up 33% of the population, with 95% credible intervals of 28–37%. The upper bound of the 95% credible interval for TWC VII is approximately 1%, suggesting that the number of animals in this oldest TWC represents a very small proportion of the total population (Fig. 3.5).

Using proportions

One of the advantages of WinBUGS is that it is easy to calculate arbitrarily complex functions of parameters and propagate the uncertainty in those parameters through the calculations when making predictions. For example, the proportion of individuals in a community that belong to

Table 3.1. *Tooth wear class of female koalas at Snake Island in 1997 (data from McLean, 2003). The tooth wear classes are ordered from youngest (TWC I) to oldest (TWC VII).*

TWC	I	II	III	IVA	IVB	IVC	V	VI	VII
Frequency	55	132	88	48	31	26	14	3	0

Box 3.14
Analysing multinomial proportions

McLean (2003) classified koalas into one of nine possible tooth wear classes. A useful model for this kind of data is a multinomial distribution, where the chance of an individual being classified into a given class is a specified probability. The sum of these probabilities must be equal to one, because the classes define all possible classifications. The data are used to estimate these probabilities, which reflect the age structure of the population.

The Dirichlet distribution is a useful prior for the probabilities because it is a multivariate distribution (with one value for each class) and the sum of the probabilities is equal to one, as required for the multinomial analysis. It is the multivariate equivalent of the beta distribution. An uninformative Dirichlet distribution can be specified by setting all its parameters equal to one. See Appendix B for more information on the Dirichlet, multinomial, and other distributions.

In the analysis of the data on the koala age structure (Table 3.1), we will use an uninformative prior for the age structure. The WinBUGS code may be written as:

```
model
{
  Y[1:9] ~ dmulti(p[1:9], N)   # N is equal to sum(Y[])
  # number of koalas in each tooth wear class drawn
    from a multinomial distribution
  p[1:9] ~ ddirch(alpha[])
  # uninformative prior for proportions (p[]) if all
    values of alpha are equal to one
}
```

The data may be entered as:

```
list(N = 397, Y = c(55, 132, 88, 48, 31, 26, 14, 3, 0),
alpha = c(1, 1, 1, 1, 1, 1, 1, 1, 1))
```

The result of taking 100 000 samples from the posterior distribution provides the predicted age structure of the koala population on Snake Island (Fig. 3.5).

different species is used to calculate diversity indices. For example, Shannon's diversity index is (Begon *et al.*, 2005):

$$H = -\sum_{i=1}^{S} P_i \ln(P_i),$$

where S is the number of species in the community and P_i is the proportion of individuals in the community that belong to species i. We will never know the proportions precisely, so our estimate of Shannon's diversity index will be imprecise. Although it might be possible to propagate the uncertainty in some simple data transformations or by using re-sampling methods in a frequentist framework, this example is easy to analyse with Bayesian methods (Box 3.15).

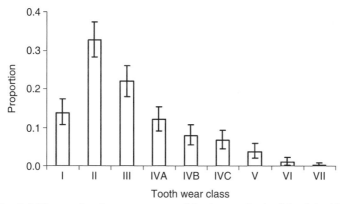

Fig. 3.5 The predicted age structure of koalas on Snake Island in 1997 (McLean, 2003). The columns represent the means of the posterior distribution and the error bars represent 95% credible intervals.

Box 3.15
Uncertainty in complex functions—diversity of a pond community

Fowler *et al.* (1998) describe a sample of 75 pupae of *Dixella* species that were obtained from a pond.

D. autumnalis	D. aestivalis	D. amphibia	D. attica
24	32	10	9

The abundance of the different species in the sample can estimate the proportion of each in the community. As in Box 3.14 for the assignment of koalas to tooth-wear classes, estimates of the proportions are uncertain. This uncertainty can then be propagated through to our estimate of species diversity. The WinBUGS model is essentially the same as for the koala example, but with an additional calculation of the diversity index.

```
model
{
  N <- sum(f[1:4])                # N equals 75 in this
                                    example
  p[1:4] ~ ddirch(alpha[])        # proportion of each
                                    species in the
                                    community
  f[1:4] ~ dmulti(p[], N)         # number of each species
                                    in the sample
  for (i in 1:4)                  # for each of the 4 species
  {
    h[i] <- p[i]*log(p[i])        # individual components
                                    of Shannon's index
  }
  H <- -sum(h[1:4])               # Shannon's index is the
                                    sum of the
                                    4 components
}
list(alpha = c(1,1,1,1), f = c(24, 32, 10, 9))
```

The mean of the posterior distribution for the Shannon diversity index is 1.25 and the 95% credible interval is 1.13–1.34. A frequentist approach would require a re-sampling method (e.g. the jack-knife) to estimate the confidence interval. For comparison, the jack-knife method produces a standard error of 0.059 and a 95% confidence interval of 1.13–1.37, which is similar to the 95% credible interval.

Concluding remarks

This chapter introduced methods for analysing averages and frequencies that are based on models in which the observations are drawn from a probability distribution. The normal, Poisson, binomial, and multi-nomial distributions were introduced for these purposes. By using prior information, Bayesian methods improve the precision of parameter estimates, and uncertainty in parameter estimates can be easily propagated in calculations that use them.

A further advantage of using Bayesian methods is that hierarchical models, in which some parameters of the model are treated as a random variable, are relatively easy to calculate and compare to equivalent non-hierarchical models. Before extending these analyses of means and frequencies to explore how they might vary in relation to explanatory variables, it is necessary to describe how the adequacy of models can be evaluated. This is examined in the next chapter, drawing on some of the examples from this chapter.

4

How good are the models?

Statistical analyses rely on models that approximate reality. For example, when analysing a mean, it is usually assumed that the data are drawn from a particular probability distribution. Such probability distributions are unlikely to emulate perfectly the real processes that generated the data. Because models are imperfect, ecologists need to consider both how well a model approximates reality, and also how well different models perform relative to each other.

There is a range of approaches to evaluating data and models. Exploratory data analysis uses graphs to detect outliers and errors, and illustrate trends prior to formal analysis (Tukey, 1997; Ellison, 2001; Quinn and Keough, 2002; Gotelli and Ellison, 2004). As part of the formal analysis, it is necessary to assess the various assumptions of the statistical model that is being used, and how well the model fits the data. Analyses of residuals, influence diagrams, probability plots, and other diagnostic methods are available for assessing the model's assumptions (Sokal and Rohlf, 1995; Quinn and Keough, 2002; Gotelli and Ellison, 2004). R-squared (R^2) values are commonly used by ecologists to determine how well a statistical model fits the data, but other measures are used for some types of analyses (e.g. ROC, Kappa values, probability plots, Quinn and Keough, 2002; Wintle et al., 2005b). All these aspects of model evaluation are important, but I do not deal with them here because they are not uniquely Bayesian.

This chapter focuses on methods of evaluating models that are unique to Bayesian statistics. The principal questions are which model is the best of several competing models, and whether there are some models that appear to be almost as good as the apparent best model (Hilborn and Mangel, 1997). The emphasis is on which models are the best candidates of those that are available.

94

There are two main features that need to be considered when evaluating the relative performance of models. The first is how well the different models fit the data. A well-fitting model will describe both the central tendency of the data and the variation in the data, and will make both unbiased and precise predictions. The second aspect that needs to be considered when evaluating models is parsimony, i.e. all else being equal, a simple model is better than a more complicated model. This is because the aim of modelling is to provide a simplified representation of reality − we do not want to include unnecessary complexity.

These two aspects (fit and simplicity) are desirable features of models, but they are also somewhat antagonistic. It is always possible to get a better fit by adding extra complexity to a model. However, there is a point where the extra complexity only provides a small improvement in the fit of the model. More importantly, there will be a point at which extra complexity, while improving the fit of the model to the data, will actually reduce the model's predictive performance. Fitting more parameters increases the possibility of generating spurious associations simply by chance. How do we determine whether the extra complexity is warranted? Below I discuss the two aspects of model fit and simplicity separately and then discuss a way of combining them to provide an overall evaluation of a model.

How good is the fit?

In any modelling exercise, it is a good idea to plot the data. In particular, comparing the predictions and observations graphically can illustrate where a model could be improved. For example, McCarthy *et al.* (1994) modelled the number of young raised by helmeted honeyeater pairs as a Poisson distribution. The parameter for the Poisson distribution can be estimated by assuming an uninformative prior, and the predicted probability distribution can be compared to the observed distribution (Box 4.1). There is reasonably good agreement between the observations and the predictions when using the mean of the posterior distribution for the Poisson parameter (Fig. 4.1).

One might also ask whether some of the differences between the predictions and the observations are unusually large. It is possible to calculate credible intervals for the observed proportions and compare the predicted probabilities to those intervals. The credible intervals on the observed proportions can be calculated by using a multinomial distribution (Box 4.2).

Box 4.1
Analysing the mean of a Poisson distribution, using a gamma prior

The helmeted honeyeater is a rare passerine occurring to the east of Melbourne in southeastern Australia. Modelling the annual number of young raised by pairs of this bird is straightforward using a Poisson distribution in WinBUGS. We simply loop over the 35 pairs for which data are available, and estimate the parameter of the Poisson distribution. It is the same as the model in Box 3.4, but uses an uninformative gamma distribution (see section on conjugacy in Appendix B) as the prior for the parameter of the Poisson distribution.

```
model
{
  lambda ~ dgamma(0.001, 0.001)        # broad prior
                                         for mean
                                         productivity
  for (i in 1:35)                      # for each of
                                         the 35 pairs

  {
    Offspring[i] ~ dpois (lambda)      # productivity
                                         drawn from a
                                         Poisson
                                         dist'n

  }
}
```

The data are simply the number of offspring raised by each pair in a breeding season:

```
list(Offspring  =  c(0,0,0,0,0,1,1,1,1,1,1,1,1,1,
2,2,2,2,2,2,2,2, 3,3, 5, 0,0,0,0,0, 1,1,1,1, 2))
```

If the model is an accurate description of the data, most if not all of the 95% credible intervals will encompass the predictions of the Poisson distribution. Given that the predictions are generally well within the credible intervals, there is little to suggest that the fit of the model is poor (Fig. 4.1). We will return to a more formal comparison of the predictions and observations in this example later in the chapter when we consider both the fit and simplicity of a model simultaneously.

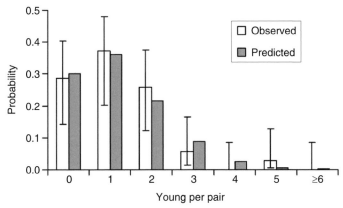

Fig. 4.1 Predicted and observed probability distributions of the number of young raised by helmeted honeyeater pairs. The prediction is based on a Poisson distribution with a mean of 1.2. The calculation of the 95% credible intervals on the observed probabilities (bars) is described in Box 4.2.

Box 4.2
Credible intervals for proportions

A method for calculating credible intervals for proportions is described in Box 3.9, and this could be used to construct such intervals for the probability distribution of helmeted honeyeater productivity. An alternative method for achieving this task is described here to illustrate the use of the categorical distribution dcat().

It is possible to construct any arbitrary probability distribution in WinBUGS by using the dcat() distribution. The required parameter for dcat() is an array of probability values. The only requirement (as for any probability distribution) is that these probabilities sum to one. In using the categorical distribution, we assume that the outcome (number of young raised) of each pair occurs with a particular probability.

Five different results have been observed for the productivity of helmeted honeyeaters (0, 1, 2, 3, and 5 offspring per pair). Another two would be possible (4, and ≥ 6). Thus, there are 7 possible classes for the observed data.

The Dirichlet distribution is a useful prior for the probabilities of the categorical distribution because it ensures that their sum is one (see Appendix B). An uninformative Dirichlet distribution has all the parameters set to values of 1. The following code could be used to implement this model in WinBUGS.

```
model
{
  for (i in 1:35) # for each of the 35 pairs
  {
    OClass[i] <- Offspring[i] + 1   # recode the data
                                      (0->1, 1->2,
                                      etc)
    OClass[i] ~ dcat(p[1:7])        # each outcome is
                                      drawn    from    a
                                      categorical
                                      dist'n
  # recoding is necessary b/c the smallest number
    returned from dcat() is 1, not 0
  }
  p[1:7] ~ ddirch(alpha[1:7])       # the prior for the
                                      prob's of each
                                      category
}
```

And the data are given by:

```
list(Offspring   =   c(0,0,0,0,0,1,1,1,1,1,1,1,1,1,
2,2,2,2,2,2,2,2, 3,3, 5, 0,0,0,0,0, 1,1,1,1, 2),
alpha = c(1,1,1,1,1,1,1))
```

The variable Offspring[] records the number of young raised by each of the 35 pairs. Note that this value is incremented by one in the code so that it refers to the appropriate probability. This is necessary because WinBUGS does not permit p[0] to be used. Arrays in WinBUGS are indexed from 1, so p[1] represents the probability of a pair raising no young, p[2] represents the probability of raising one young, etc. Thus, p[i] provides the probability of raising i-1 young, except with p[7] being the probability of raising six or more young. A total of 100 000 samples from the posterior distribution leads to the 95% credible intervals shown in Fig. 4.1.

A measure of fit

The above graphical procedure is only one of many for evaluating the fit of a model. Although these graphical methods are useful, this section describes a quantitative approach to evaluating the fit of a model. This approach helps to formalize the process of model evaluation and allows different models to be compared. So while the Poisson distribution seems to provide a reasonable fit (Fig. 4.1), an alternative model may do better or worse.

In the predictions in Fig. 4.1, it was assumed that the number of offspring raised by a pair was drawn from a Poisson distribution. For this distribution, the probability of the number of offspring equalling x is given by the formula:

$$\Pr(X = x) = e^{-\lambda}\lambda^{x}/x!,$$

where λ is the parameter of the probability distribution (see Box 3.3).

This formula can be used to calculate the likelihood of obtaining a particular result for different values of λ. For the sake of illustration, consider a situation where only two observations are made of helmeted honeyeater productivity, with two and three offspring being observed. If the two observations are assumed to be independent, then the joint likelihood of these two observations is $L = (e^{-\lambda}\lambda^{2}/2!) \times (e^{-\lambda}\lambda^{3}/3!)$. The value of λ that maximizes the likelihood L of obtaining the data is the maximum likelihood estimate, which is $\lambda = 2.5$ in this case (Fig 4.2a). The likelihood at this value of lambda and for these two observations is 0.0548 $[L_{max} = (e^{-2.5}2.5^{2}/2!) \times (e^{-2.5}2.5^{3}/3!)]$.

An alternative model would be to permit different values of lambda for each observation. Unsurprisingly, the maximum likelihood for this alternative model is obtained when lambda equals two for the observation of two, and three for the observation of three. This model is more complex because it has two parameters rather than one. The maximum value for the likelihood is 0.0606 $[L_{max} = (e^{-2}2^{2}/2!) \times (e^{-3}3^{3}/3!)]$. Therefore, we would conclude that the second model provides a better fit to the data than the first because its likelihood is greater. In fact, it would be impossible for this second model to fit worse than the first because it contains greater flexibility by virtue of its extra parameter.

The approach illustrated above with a sample size of two can be extended to any sample size, by simply multiplying together the likelihoods for all the individual observations. To obtain the likelihood for the 35 observations of pair productivity, we simply calculate the likelihood for

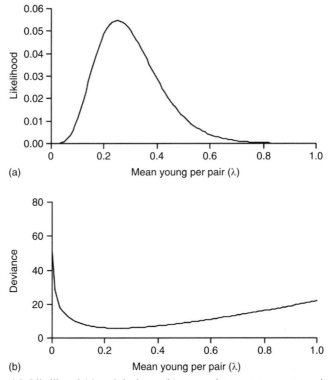

Fig. 4.2 Likelihood (a) and deviance (b) versus the mean young per pair for the Poisson model with data from two pairs, one of which raised two young and the other three. The same parameter value that maximizes the likelihood (2.5) also minimizes the deviance, with the minimum deviance (D_{min}) equal to 5.81.

each pair and multiply them together, assuming that the data have arisen independently. For larger datasets, the likelihood value will tend to decrease towards zero. For example, the maximum likelihood for the data on the productivity of the 35 helmeted honeyeater pairs ($L = [e^{-\lambda}\lambda^0/0!]^{10}[e^{-\lambda}\lambda^1/1!]^{13}[e^{-\lambda}\lambda^2/2!]^9[e^{-\lambda}\lambda^3/3!]^2[e^{-\lambda}\lambda^5/5!]$) is approximately 5.5×10^{-22}, which is obtained when $\lambda = 1.2$.

Taking the natural logarithm of the likelihood helps to resolve numerically small differences that may nevertheless be important when the likelihood is close to zero. Thus, for the first model considered above and the two observations of 2 and 3, the maximum log-likelihood is -2.903, and for the second (more complex) model, it is -2.803.

For several reasons that I will not elaborate (see Edwards, 1992, Hilborn and Mangel, 1997, Burnham and Anderson, 2002 for further information), log-likelihood values are often converted to what is known as the 'deviance' (D) of the model by multiplying by -2. Therefore, the minimum deviance (D_{min}) is 5.81 for the first model (the smallest value in Fig. 4.2b) and 5.61 for the second. A larger value for the deviance indicates a poorer fit. The deviance is used as a basis for comparing models in a range of statistical approaches.

In maximum likelihood estimation, the best parameter estimates are those that maximize the likelihood. The same parameter values that maximize the likelihood also minimize the deviance (Fig. 4.2) because there is a negative monotonic relationship between the two ($D = -2\ln(L)$). The parameter estimates that maximize the likelihood are equivalent to the mode of the posterior distribution in a Bayesian analysis that employs uninformative priors.

The deviance calculated at the mode of the posterior will be the minimum deviance (D_{min}). If the posterior distributions of the parameters of a model are symmetrical, the mode and the mean will be equal. Therefore, the deviance of the model calculated at the mean of the posteriors ($\hat{D}$, called 'D hat') will approximately equal D_{min}. The calculation of the deviance at the mean of the posterior distributions is automated for many models in WinBUGS, making it easy to calculate. The procedure is described in Box 4.3 for the models in Boxes 4.1 and 4.2.

Quantitative measures of model fit help to formalize the process of model evaluation. Such a measure is provided by the deviance, with lower values indicating a better fit. If calculated at the 'best' parameter estimates, it can be used to measure the relative fit of different models. More complex models have greater ability to conform to the data, so they tend to fit better. Whether the better fit provided by a more complex model is justified is another matter, which will be dealt with after considering how to measure model complexity.

How complex is the model?

The Poisson model of the number of young raised by helmeted honeyeater pairs is very simple (Box 4.1). Its simplicity is embodied in the single parameter that needs to be estimated. The model based on a categorical distribution is more complex, requiring seven parameters to be estimated (Box 4.2). A key question is whether the extra parameters are justified. We will make this comparison more formally in the next section where the trade-off between model simplicity and model fit is considered.

Box 4.3
Calculating $\hat{D}$ as a measure of model fit

The deviance calculated at the mean of the posterior distribution
is generated automatically in WinBUGS for many statistical
models. To calculate this value, take samples until the stationary
distribution is reached and then open the DIC Tool (under the
Inference menu). Click 'set' on the DIC Tool, and then generate
numerous samples from the posterior distribution. Click on
'DIC' in the DIC Tool, and the results of the analysis will be
displayed. The deviance at the mean of the posterior distribution ($\hat{D}$)
for different elements of the data will be displayed under the column
title Dhat.

For example, consider the model of helmeted honeyeater
productivity in Box 4.1. After excluding the first 10 000 samples as a
burn-in, clicking 'set' in the DIC Tool, and taking a subsequent
100 000 samples, the following output is obtained by clicking 'DIC' in
the DIC Tool.

	Dbar	Dhat	DIC	pD
Offspring	98.91	97.9	99.91	1.004
total	98.91	97.9	99.91	1.004

The rows provide the results for each individual set of data
(Offspring in this case) and for all the data combined (total),
which in this case is equivalent to offspring because there is only
one set of data. The deviance at the mean of the posterior
distribution ($\hat{D}$, Dhat) is 97.9. Dbar, pD, and DIC are explained later
in the chapter.

For the model that uses the categorical distribution (Box 4.2), the
equivalent output from WinBUGS is:

	Dbar	Dhat	DIC	pD
OClass	101.0	97.83	104.2	3.184
total	101.0	97.83	104.2	3.184

Therefore, this second model provides a better fit because the deviance at the mean of the posterior distribution ($\hat{D}$, Dhat) is smaller (97.83 versus 97.9). This is not surprising; the second model has greater flexibility to fit to the data because the distribution is not constrained to be a Poisson.

However, the complexity of a model cannot always be measured by simply counting the number of parameters, particularly when including random effects or informative prior distributions. For example, consider the example where the mean density of oaks varied among quadrats (Box 3.5). In this case, each quadrat has its own mean density, so we might think that the number of parameters is equal to the number of quadrats (ten in this case). However, the means of the quadrats are not completely independent of each other, being drawn from the same probability distribution. The dependence among the parameters means that the effective number of parameters will be less than ten.

The effective number of parameters will also be less than the actual number of parameters when the prior distributions are informative. When a parameter has an informative distribution, it is not free to vary entirely over its possible range. In the extreme case where the prior is very precise, it is effectively fixed as a constant and makes little contribution to the effective number of parameters. In the other extreme, where the prior is uninformative, the parameter is free to vary across a wide range and only the data are available to estimate it. In this case, its estimation contributes a full parameter to the effective number of parameters. For intermediate cases, where the prior is somewhat informative, the parameter contributes proportionally to the effective number of parameters.

Spiegelhalter *et al.* (2002) proposed a measure of the effective number of parameters. A brief intuitive description is provided here, although readers are encouraged to read the article by Spiegelhalter *et al.* (2002) for themselves. This measure of the effective number of parameters (p_D) is obtained as the difference between the mean of the posterior deviance of the model ($\bar{D}$, 'D-bar') and the deviance of the model when using the means of the posterior distributions of the parameters ($\hat{D}$, which is the previously mentioned measure of fit):

$$p_D = \bar{D} - \hat{D}.$$

Because the deviance is approximately minimized when the parameters are equal to their posterior means, different values of the parameters will lead to larger values of the deviance. If the parameters are sampled from their posterior distributions and the deviance is recalculated for each sample, the mean of these recalculated deviances is the mean of the posterior deviance ($\bar{D}$).

We would expect the mean of the posterior deviance ($\bar{D}$) to be greater than the minimum deviance (which is approximately $\hat{D}$). The more parameters there are, the greater the scope for differences from the minimum (i.e. it is less likely that all the parameters are sampled close to the values that generate the minimum deviance). However, if a parameter is somewhat constrained (e.g. either by a prior or because it is a random effect drawn from a probability distribution it shares with other random effects), it will be less likely to take a value that is different from the one that generated the minimum deviance. Therefore, constraints on the parameters will tend to reduce the possible difference between the mean of the posterior deviance ($\bar{D}$) and the deviance when the mean parameter values are used ($\hat{D}$).

It turns out that the difference between the mean of the posterior deviance and the deviance of the model when using the means of the parameters is approximately equal to the effective number of parameters. This property can be illustrated by the model of helmeted honeyeater productivity. For the model that uses a Poisson distribution (Box 4.1), p_D equals 1.004 (obtained in WinBUGS as the value of pD from the DIC output in Box 4.3), which is essentially equal to the number of parameters being estimated (the single parameter of the Poisson distribution).

The model that uses the categorical distribution for helmeted honeyeater productivity (Box 4.2) has seven parameters (one for each outcome zero to five, and one for the outcome ≥ 6). However, these parameters are not free to vary without restraint because they must sum to precisely one. When six have been specified, the seventh is pre-determined. Therefore, there are certainly no more than six effective parameters. However, even when one parameter is defined, the maximum possible value for any of the others is limited. As an extreme example, if one of the probabilities is equal to one, then all the other probabilities must equal zero. Therefore, it would not be surprising to find that the effective number of parameters in this example is less than six. In fact, the estimate of the effective number of parameters is 3.184 in this case (Box 4.3).

Readers may wish to calculate the effective number of parameters for some of the other models developed in Chapter 3. For example, $p_D = 2.1$ for the two-parameter model of the mean difference in the size of spider webs (Box 3.4), $p_D = 1.011$ for the one-parameter model of red oak density (Box 3.4), $p_D = 6.4$ for the hierarchical model of red oak density (Box 3.5), and $p_D = 6.65$ for the categorical model of koala age structure (Box 3.9).

So far we have considered two models for helmeted honeyeater productivity; one using a Poisson distribution and one based on a more flexible categorical distribution. How do we determine which is better? Not surprisingly, the model with more parameters provided a better fit (Box 4.3). However, a simpler model is preferable to a more complex model. How do we assess the trade-off between these two measures of simplicity and fit that we have assessed separately? The answer lies in a combined measure of the two, in which the fit of a model is penalized by its complexity.

Combining measures of fit and simplicity

Akaike's information criterion

By definition, models are meant to approximate reality. The best model is one that explains the maximum level of detail in the simplest possible way. Akaike (1973) considered how to measure the amount of information lost when using a model to approximate reality. The approach measures how well different models approximate reality even though reality may be unknown. Models that lose the least amount of information will tend to make the best predictions of replicate datasets.

This relative measure of the information content might be appreciated by an analogy of travel between two stars Alpha Centauri and Betalgeuse. If we were told we had travelled for 1000 km towards Betalgeuse, we would know we were 1000 km closer to our destination, but we would not necessarily know how much further there was to go. When using Akaike's approach to evaluate the relative information content of models, we measure how much closer each model is to reality, not the actual distance between the models and reality. This latter quantity can be evaluated using other measures of fit such as R^2 values, predictive performance, etc.

Akaike demonstrated a relationship between the expected information content of a model and the log-likelihood at its maximum point (i.e. the minimum deviance). Akaike (1973) estimated the relative amount of information that is lost when using models to describe the truth. This estimate of the information loss leads to Akaike's Information Criterion, which is given by the formula:

$$\text{AIC} = D_{\min} + 2K,$$

where $D_{\min}$ is the smallest deviance for the model (i.e. the deviance when using the best-fitting parameter set for the model) and K is the number of estimated parameters. We have encountered the two parts of this formula in the two previous sections of this chapter. When deciding between two or more models, the best model can be chosen as the one with the lowest AIC value, because it is expected to lose the least amount of information. Put another way, the model with the smallest AIC is expected to provide the best predictions for a replicate set of data.

Deviance information criterion

Spiegelhalter *et al.* (2002) proposed an alternative to AIC, known as the deviance information criterion (DIC). This has a very similar form to AIC and is given by:

$$\text{DIC} = \hat{D} + 2p_D,$$

where $\hat{D}$ is the deviance when using the mean of the posterior distributions for the parameters and p_D is the effective number of estimated parameters. The similarities between AIC and DIC are evident. $\hat{D}$ will be equal to $D_{\min}$ when the posterior distributions are symmetrical. Further, p_D and K will be approximately equal for models without constraints on parameters.

As noted in the previous section, it can sometimes be difficult to determine the effective number of parameters in a model (K), especially when random effect terms and informative prior distributions are used. However, the effective number is easily estimated as p_D with MCMC methods, making DIC valuable for comparing Bayesian models.

How different are the models?

It is worth remembering that AIC and DIC only *estimate* the relative information content of models. The values of AIC and DIC will depend on the particular data that were collected. Burnham and Anderson (2002)

Table 4.1. *Interpretation of the level of support for apparently inferior models relative to the model with the lowest AIC, based on differences in their AIC values (from Burnham and Anderson, 2002, p. 70).*

ΔAIC	Degree of support
0–2	Substantial
4–7	Considerably less
> 10	Essentially none

developed rules of thumb for assessing differences in AIC values between a given model and the model with the smallest AIC. These differences are given by ΔAIC (Table 4.1). A model with an AIC value within 2 units of the smallest AIC has substantial support. Larger differences suggest less support.

Similar interpretations apply for DIC values (Spiegelhalter *et al.*, 2002). Therefore, Table 4.1 can be used to help determine the set of plausible models. Those with DIC values within 10 of the smallest DIC value might be regarded as possibly the best model, while those within 2–4 might be regarded as the more likely candidates.

Consider the two models of helmeted honeyeater productivity. The DIC for the Poisson model (Box 4.1) was 99.9 and the categorical model (Box 4.2) was 104.1 (obtained from the DIC output in WinBUGS, Box 4.3). The difference (4.2) suggests there is relatively strong support for the first model relative to the second (Table 4.1), but it does not mean that the first model is good in an absolute sense. This latter attribute can be assessed by the fit (e.g. Fig. 4.1).

The model of the number of young raised by helmeted honeyeater pairs based on a Poisson distribution (Box 4.1) ignores almost everything that we know about the productivity of birds by assuming that all pairs have the same average productivity. However, pairs that remain intact for the entire breeding season have a greater opportunity to produce more offspring than pairs that do not remain intact. The main reason that pairs do not remain intact is the death of one or both of the birds.

A model that accommodates differences in productivity for the two types of pairs would require two parameters; one for the pairs that remain intact for the breeding season and one for the pairs that split. Because this model has two parameters, it is more complex than the model with a single parameter in which the mean productivity of all pairs was the same. A key question is whether this extra parameter is justified. Given that the mean number of offspring raised by split pairs is 0.6 (range 0–2), while that of intact pairs is 1.44 (range 0–5), we might think that it is. The DIC value for the two parameter model (DIC = 97.2, Box 4.4) is smaller than the single parameter model (Box 4.1) suggesting that the extra parameter is warranted, although the evidence is not compelling. It makes biological sense that pairs that breed for longer will produce more offspring on average than those that do not. Therefore, it would be reasonable to accept the more complex model as the better of these two.

Assessing priors

DIC values are most commonly interpreted as measuring the relative performance of different models, because they arise from the literature of model evaluation. However, different priors can also be compared using DIC values. These priors might represent different points of view, and when confronted with the same data and model we can evaluate which views are most consistent with the evidence. An example is given in Chapter 6.

The Bayes factor and model probabilities

Information criteria such as DIC and AIC are used to help select among competing models. Posterior model probabilities based on Bayes factors (Jeffreys, 1961; Kass and Raftery, 1995) offer an alternative approach by evaluating the probability that the different models are correct. Of course, such a definitive statement requires that the set of models is exhaustive. If it is not, then the Bayes factor will only provide the relative probability of those models being considered (Link and Barker, 2006).

The posterior probability that a particular model (M_i) is correct (given the data D) is (Jeffreys, 1961):

$$\Pr(M_i \mid D) = \frac{\Pr(M_i)\Pr(D \mid M_i)}{\sum_j \Pr(M_j)\Pr(D \mid M_j)}.$$

This is a simple application of Bayes' rule, with the denominator being a sum over all the models being considered. The values $\Pr(D \mid M_j)$ are the prior probabilities of obtaining the data under each of the different models M_j.

Box 4.4
Comparing different models

The productivity of helmeted honeyeater pairs is likely to depend on whether the pair remains intact over the entire breeding season. This can be specified in WinBUGS by having two different parameters for the Poisson distribution, one for the pairs that remain intact and one for the others. It is possible to include a variable in the data that identifies which of the pairs remained intact. This variable can then be used to determine which of the two means is used. The WinBUGS code would then be:

```
model
{
   lambda[1] ~ dgamma(0.001, 0.001)  # mean if pairs
                                          split
   lambda[2] ~ dgamma(0.001, 0.001)  # mean if pairs
                                          remain intact
   for (i in 1:35)                    # for each of the
                                          35 pairs
   {
      Offspring[i] ~ dpois(lambda[Intact[i] + 1])
      # actual offspring drawn from Poisson
      # mean (lambda) depends on whether pair remains
        intact or not
   }
}
```

We also need to include the variable `Intact` to discriminate between pairs that remain intact (taking a value of 1), and those that do not (taking a value of 0):

```
list(Offspring  =  c(0,0,0,0,0,1,1,1,1,1,1,1,1,1,
2,2,2,2,2,2,2,2, 3,3, 5, 0,0,0,0,0, 1,1,1,1, 2),
Intact = c(1,1,1,1,1,1,1,1,1,1,1,1,1,1,1,1,1,1,1,
1,1,1,1,1,1, 0,0,0,0,0,0,0,0,0,0))
```

The DIC value for this model is 97.2, which suggests it is a better model than the one that ignored differences in productivity among pairs (Box 4.1, DIC = 99.9), but the difference is not large. As expected, the estimated number of parameters is approximately two ($p_D = 2.03$). The fit of the model appears to be good, with all predictions contained within the 95% credible intervals around the observed proportions (Fig. 4.3).

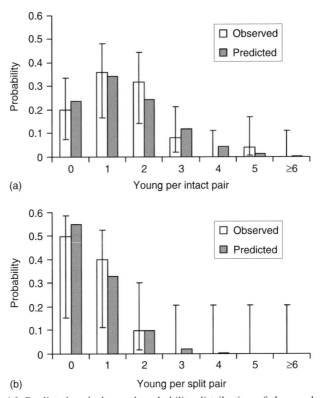

(a)   Young per intact pair

(b)   Young per split pair

Fig. 4.3 Predicted and observed probability distribution of the number of young raised by intact and split helmeted honeyeater pairs. The 95% intervals on the observed probabilities (bars) were calculated using the method described in Box 4.2. The predictions are based on two different Poisson distributions with means of 1.44 and 0.6 respectively.

For the sake of illustration, only two models will be considered, M_1 and M_2. If we consider the ratio of the posterior model probabilities $\Pr(M_1 \mid D)$ and $\Pr(M_2 \mid D)$, the denominator cancels out of the above equation, leading to:

$$\frac{\Pr(M_1 \mid D)}{\Pr(M_2 \mid D)} = \frac{\Pr(M_1)\Pr(D \mid M_1)}{\Pr(M_2)\Pr(D \mid M_2)} = \frac{\Pr(M_1)}{\Pr(M_2)} \times \frac{\Pr(D \mid M_1)}{\Pr(D \mid M_2)}.$$

The latter ratio $(\Pr(D \mid M_1)/\Pr(D \mid M_2))$ is the Bayes factor, and is commonly represented by the symbol B. It is the degree to which the ratio of the model probabilities (the model odds) changes when the data are considered. If the Bayes factor $B > 1$, the first model (M_1) is supported by the data more than M_2. M_2 is favoured over M_1 when $B < 1$.

The Bayes factor is similar in form to the likelihood ratio. The difference is that the likelihood ratio is calculated at the particular values of the parameters (those that maximize the likelihood), while the Bayes factor is based on the prior probability of the data, so it is integrated over the values of the parameters specified by the prior.

Bayes factors are conceptually simple, but there are two main difficulties with their use. Firstly, it can be difficult to calculate the values $Pr(D \mid M_j)$. A simple example is illustrated below, which is hard enough. There are numerical methods for more complex problems (e.g. see Kass and Raftery, 1995). The other difficulty is that the Bayes factor can be very sensitive to the choice of the prior. The choice of prior often has little influence when estimating parameters and constructing credible intervals, but it can be very influential when calculating Bayes factors.

The difficulty in calculating the values $Pr(D \mid M_j)$ arises because the probability is integrated over possible values of the models' parameters. For example, consider models of sex ratio. The first model (M_1) might be that the proportion of males and females are equal, so the proportion of males (p) is equal to 0.5. If there are 20 males and 8 females observed in a sample of deer, for example, (Flueck, 2001), it is easy to calculate the probability of the data for M_1 using binomial probabilities (Box 2.1), which is 0.0116.

The alternative model (M_2) might be that the sex ratio is not equal to 0.5. However, this alternative model needs to be specified in terms of a probability distribution for p the proportion of males. If any value between zero and one was considered equally likely, a priori, then the appropriate distribution would be uniform between zero and one. For such a distribution, the probability density is given by $f(p) = 1$.

The probability of the data under this model (0.0345, Box 4.5) allows us to calculate the Bayes factor for the two models, which is $B = 0.336$. This means the data support M_2 more than M_1. If both models had equal prior probabilities of 0.5, the posterior probabilities of the two models would be $Pr(M_1 \mid D) = 0.25$ and $Pr(M_2 \mid D) = 0.75$. Therefore, there is not compelling evidence that the sex ratio p is different from 0.5, but the data are suggestive. This is a form of hypothesis testing because we have evaluated the probability that the sex ratio p is 0.5 versus the alternative that $p \neq 0.5$. However, Bayesian hypothesis testing is different from usual null hypothesis testing because the calculation is not influenced by more extreme data than those observed, and it provides direct statements about the probability of the different hypotheses being correct, rather than focusing on the probability of obtaining the data.

Box 4.5
Model probabilities and Bayes factors

Flueck (2001) determined the sex of 28 deer to determine if the sex ratio differed from parity. The first model can be expressed as the sex ratio p is equal to 0.5. Under this model the probability of obtaining the data (20 males) can be calculated as:

$$\Pr(D \mid M_1) = \Pr(20 \text{ males \& 8 females} \mid p = 0.5)$$

$$= \left(\frac{28!}{20!8!}\right) 0.5^{20} (1 - 0.5)^8 = 0.0116.$$

However, when calculating the probability of the data for the model in which the sex ratio (p) is not equal to 0.5 (M_2), we need to weight p by the prior probabilities and integrate over the possible values:

$$\Pr(D \mid M_2) = \Pr(20 \text{ males \& 8 females} \mid p \text{ is not equal to } 0.5)$$

$$= \int_0^1 \left(\frac{28!}{20!8!}\right) p^{20} (1 - p)^8 f(p) dp,$$

$$= \left(\frac{28!}{20!8!}\right) \int_0^1 p^{20} (1 - p)^8 f(p) dp.$$

where $f(p)$ is the prior probability distribution for the sex ratio p when it is not equal to 0.5.

One possible prior would be that the sex ratio is uniformly distributed between zero and one, so the prior probability density function is given by $f(p) = 1$ (Appendix B). Thus:

$$\Pr(D \mid M_2) = \left(\frac{28!}{20!8!}\right) \int_0^1 p^{20} (1 - p)^8 dp.$$

Some tedious algebra provides the solution of this integral, which leads to:

$$\Pr(D \mid M_2) = \left(\frac{28!}{20!8!}\right) \frac{1}{90135045} = \frac{3108105}{90135045} = 0.0345.$$

Therefore, the Bayes factor is equal to:

$$B = \Pr(D \mid M_1) / \Pr(D \mid M_2) = 0.0116/0.0345 = 0.336.$$

If both models had equal prior probabilities of 0.5, the posterior probabilities of the two models would be:

$\Pr(M_1 \mid D) = 0.5 \times 0.0116/(0.5 \times 0.0116 + 0.5 \times 0.0345) = 0.25$, and
$\Pr(M_2 \mid D) = 0.5 \times 0.0345/(0.5 \times 0.0116 + 0.5 \times 0.0345) = 0.75$.

Therefore, there is not compelling evidence that the sex ratio p is different from 0.5, but the data are suggestive.

A uniform prior probability for the alternative model (M_2) means that sex ratios close to zero or one are as equally likely as a sex ratio equal to 0.5. This seems improbable, with sex ratios close to, but different from, 0.5 more likely. In this case, a beta distribution centred on 0.5 could be used as the prior. Beta distributions are given by the following equation when the mean (and mode) are equal to 0.5 (Appendix B):

$$f(p) = \frac{\Gamma(2a)}{\Gamma(a)^2} p^{a-1}(1 - p)^{a-1}.$$

When the parameter a is equal to 1, the prior is a uniform distribution between zero and 1, as was evaluated previously. The precision of this prior distribution increases as a increases, with more of the probability concentrated around 0.5 (Fig. 4.4). When a is very large, the prior distribution for M_2 approaches that of M_1 ($p = 0.5$).

Using the beta prior with a mean of 0.5,

$$\Pr(D \mid M_2) = \left(\frac{28!}{20!8!}\right) \int_0^1 p^{20}(1 - p)^8 \frac{\Gamma(2a)}{\Gamma(a)^2} p^{a-1}(1 - p)^{a-1} dp$$

$$= \left(\frac{28!}{20!8!}\right) \frac{\Gamma(2a)}{\Gamma(a)^2} \int_0^1 p^{20+a-1}(1 - p)^{8+a-1} dp$$

This can be solved for a given value of a by using the definition of the beta function (Appendix B, p. 267):

$$\int_0^1 t^{a-1}(1 - t)^{b-1} dt = \frac{\Gamma(a)\Gamma(b)}{\Gamma(a + b)},$$

so $\displaystyle \int_0^1 p^{20+a-1}(1 - p)^{8+a-1} dp = \frac{\Gamma(a + 20)\Gamma(a + 8)}{\Gamma(2a + 28)}$, and

$$\Pr(D \mid M_2) = \left(\frac{28!}{20!8!}\right) \frac{\Gamma(2a)}{\Gamma(a)^2} \frac{\Gamma(a+20)\Gamma(a+8)}{\Gamma(2a+28)}.$$

The posterior probability of M_1 being correct ($\Pr(M_1 \mid D)$) can be calculated for different values of a by using this expression for $\Pr(D \mid M_2)$. This demonstrates that the smallest possible posterior probability of M_1 being correct is approximately 0.2 when a is approximately 3 (Fig. 4.5). As a increases towards infinity, the posterior probability of the models approach 0.5, which is not surprising because the two models are largely indistinguishable with high values of a. Therefore, regardless of the choice of a, an observation of 20 males in a sample size of 28 deer provides only modest evidence that the sex ratio is different from 0.5.

Note that a null hypothesis significance test of these data leads to a p-value of 0.036 for the null hypothesis $p = 0.5$ and the alternative hypothesis of $p \neq 0.5$. Using the conventional cut-off of 0.05 leads to rejection of the null hypothesis, even though the probability that the alternative is true is only 0.75.[1] This emphasizes the difference between null hypothesis significance testing and Bayesian hypothesis testing. The latter provides direct statements about the probability of hypotheses being true, while p-values tend to overstate the evidence against null hypotheses (Berger and Sellke, 1987).

A uniform prior probability for the alternative model (M_2) means that sex ratios close to zero or one are as likely as a sex ratio equal to 0.5. This seems improbable, with sex ratios close to, but different from, 0.5 more likely. In this case, a beta distribution centred on 0.5 could be used as the prior for model M_2 (Fig. 4.4). The resulting model probabilities depend on the choice of the variance for this prior. The smallest posterior probability for M_2 is 0.2, demonstrating that an observation of 20 males in a sample of 28 deer provides only modest evidence that the sex ratio is different from zero (Fig. 4.5).

Although somewhat complicated, the calculations in Box 4.5 are some of the easiest for determining Bayes factors and model probabilities.

---

[1] With Bayesian hypothesis testing there is not a conventional cut-off at which point we would reject one or more of the possible hypotheses being considered. In Bayesian analyses, this decision depends on the benefits and costs of being right and wrong, and the objective, a field of research known as Bayesian decision analysis (Gelman *et al.*, 2004).

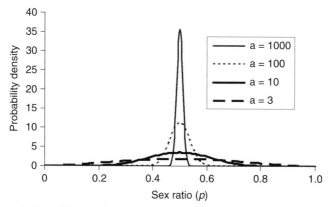

Fig. 4.4 Four different beta distributions with means of 0.5 and different variances, influenced by the parameter a.

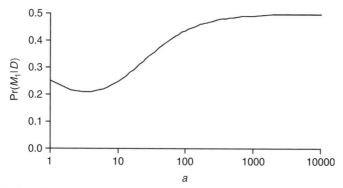

Fig. 4.5 The posterior probability that the sex ratio of deer is equal to 0.5 (M_1) for different values of a, the parameter that influences the precision of the prior distribution for the alternative model M_2. Regardless of the value of a, the posterior probability that the sex ratio equals 0.5 is always greater than 0.2.

The level of complexity for most models and the sensitivity of results to the choice of the prior mean that model probabilities are not always calculated. Information criteria, such as DIC, are commonly used as surrogates when distinguishing between competing models. However, such methods only approximate model probabilities. If ecologists wish to assign probabilities to models, they should do so by calculating model probabilities, despite the computational difficulties and sensitivity to priors.

There are a range of numerical procedures for calculating model probabilities and Bayes factors (Kass and Raftery, 1995). These analyses can be extended to consider how uncertainty in the choice of models propagates through to uncertainty in predictions and parameter estimates by using model averaging (Draper, 1995; Volinsky *et al.*, 1997; Hoeting *et al.*, 1999; Wintle *et al.*, 2003). There are methods that work in WinBUGS, by using an indicator variable to select the different models. (Carlin and Chib, 1995; see also Congdon, 2003).

Evaluating the shape of distributions

This chapter concludes by briefly describing methods for evaluating whether the shapes of distributions that are used in analyses are appropriate. In Box 3.5, it is assumed that the variables called 'mean[]' are lognormally distributed, so the variables log(mean[]) should be normally distributed. It is possible to check this, by recording their sampled values and comparing them to what would be expected from a normal distribution. This is easy to do graphically by constructing a probability plot (Chambers *et al.*, 1983).

The shape of probability distributions can also be evaluated by calculating their skewness and kurtosis. It is then possible to determine whether the posterior distribution of these statistics differ from the values expected for the given probability distribution (Gelman and Meng, 1996). Skewness measures the asymmetry of a distribution and kurtosis measures the 'fatness' of the tails. The formulae for calculating skewness and kurtosis are provided in Appendix B. For a normal distribution, skewness equals zero and kurtosis equals three, which can be compared to the values obtained from MCMC samples.

For the example in Box 3.5, the code would be modified as follows to calculate the skewness and kurtosis of the logarithm of the mean density of trees in quadrats.

```
model
{
  for (i in 1:10)
  {
    mean[i] ~ dlnorm(m, tau)
    Y[i] ~ dpois(mean[i])
    lm[i] <- log(mean[i])
    r3[i] <- pow(lm[i] - m, 3)
    r4[i] <- pow(lm[i] - m, 4)
  }
```

```
    skewness <- mean(r3[])*pow(tau, 1.5)
    kurtosis <- mean(r4[])*pow(tau, 2)
    m ~ dnorm(0, 1.0E-6)
    sd ~ dunif(0, 10)
    tau <- 1/sd/sd
}
```

The variables r3[] and r4[] measure third and fourth powers of deviations from the expected value of the normal distribution. The means of these values are used with the precision (inverse of the variance) to calculate skewness and kurtosis. For this data set, the 95% credible interval is [−2.5, 2.0] for skewness, and [0.15, 9.3] for kurtosis based on 100 000 samples from the posterior distribution. These intervals bound the expected values of zero and three, suggesting that the lognormal distribution is a reasonable model of variation in mean density of plants among quadrats, although the wide intervals suggest that other distributions might perform at least as well.

The analysis of tree diameters in Box 3.2 provides a second example of evaluating the shape of a distribution by calculating its skewness and kurtosis. The code would be modified to:

```
model
{
    mean ~ dnorm(53, 0.04)
    var ~ dlnorm(4.75, 3.05)
    prec <- 1/var
    for (i in 1:10)
    {
        Y[i] ~ dnorm(mean, prec)
        r3[i] <- pow(Y[i] - mean, 3)
        r4[i] <- pow(Y[i] - mean, 4)
    }
    skewness <- mean(r3[])*pow(prec, 1.5)
    kurtosis <- mean(r4[])*pow(prec, 2)
}
```

The resulting values suggest that the data may be positively skewed because the expected value of zero tends to occur towards the tail of the posterior distribution of the variable skewness (the 95% credibility interval of [−0.23, 4.2] only just includes zero). In comparison, the expected kurtosis is well-contained within the corresponding 95%

credibility interval of [0.89, 14.7]. Given the suggestion of positive skewness, and the constraint that diameters cannot be negative, a lognormal distribution (or equivalently log-transformation of the diameter measurements) is likely to provide a better fit in this case.

Concluding remarks

In this chapter, DIC is presented as a criterion for selecting models. Like other information criteria such as AIC and BIC (see Burnham and Anderson, 2002), it aims to assess the trade-off between fit and complexity. Good models provide a reasonable fit but are not overly complicated. AIC and its derivative DIC are based on estimating the amount of information lost when using a model to approximate reality. A model is chosen as the best if it is estimated to lose the least information; it is the model that is expected to make the best predictions for a replicate dataset.

DIC and AIC are relatively new and their application to a variety of models is not well-studied (Richards, 2005). It is also possible to calculate posterior model probabilities, i.e. the probability that a particular model is true. The required calculations can be difficult for novices, and the posterior model probabilities can be very sensitive to the choice of the prior distributions that are used for the parameters of the models. One of the main advantages of using model probabilities is that predictions can be averaged across the different competing models (Wintle *et al.*, 2003). In addition to using formal methods to assess the relative and actual performance of models, they need to be evaluated with common sense. For example, a model of the number of plants in quadrats that permits negative or fractional numbers of plants, such as a normal distribution, is inferior, in at least some respects, to a model that only permits non-negative integers, such as a Poisson or a variant of it. The selection of models to be assessed should be based on a thorough a priori assessment, so biologically reasonable models are considered.

5

Regression and correlation

Regression

In most of the previous chapters it was assumed that all samples had the same mean. While some of the models permitted the average to vary randomly among samples (e.g. Box 3.5), in general, explanatory variables were not used to describe how this mean might vary systematically among samples. One exception to this is the example in which the mean productivity of helmeted honeyeater pairs depended on whether the birds remained intact for the entire six-month breeding season (Box 4.4). In this case, whether or not the pair remained intact was the explanatory variable and the productivity was the dependent variable. Essentially, the model was a form of regression, or more precisely a form of generalized linear model. This will become clearer in this chapter. For now, it is sufficient to realize that the model consisted of a dependent variable and an explanatory variable that influenced the expected value of the dependent variable. The simplest form of this dependency is a linear relationship, which leads to simple linear regression.

Simple linear regression

In Box 3.2 it was assumed that the data were drawn from a normal distribution with a certain mean and precision. This analysis is modified in simple linear regression by modelling a linear relationship between the mean and an explanatory variable. Thus, the mean of an observation is given by an equation such as:

$$\text{mean} = a + bx,$$

where x is the value of the explanatory variable for the observation, and a and b are known as regression coefficients. The coefficient a is often referred to as the intercept, because it is the expected value of an observation when the explanatory variable is zero. The coefficient b controls the slope of the relationship between the mean and x; larger absolute values of b reflect stronger relationships between the two variables.

The details of simple linear regression can be most easily seen in an example. Consider the relationship between the abundance of coarse woody debris in lakes and the density of trees along the shoreline (Christensen *et al.*, 1996; Quinn and Keough, 2002). Lakes with greater numbers of trees would be expected to have more coarse woody debris (CWD). It turns out that such a relationship is apparent in the data (Fig. 5.1). The amount of coarse woody debris is predicted to increase by 0.115 m^2/km for every increase of one tree per km (Box 5.1). The 95% credible interval for this increase was 0.065–0.165, suggesting that we can be reasonably sure that there is a positive relationship between the two variables.

The resulting regression can use measurements of tree density to predict abundance of CWD debris. For example, if we had a lake with a density of 1500 trees per km, we could predict that the mean amount of CWD would be 95.5 m^2/km ($-77.0 + 0.115 \times 1500$). However, there is some uncertainty about the relationship between tree density and CWD, which is reflected by the credible intervals of the parameter estimates.

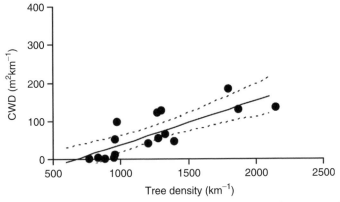

Fig. 5.1 Amount of coarse woody debris (CWD) in 16 lakes versus the density of trees along their shoreline (Christensen *et al.*, 1996). The solid line is the fitted linear regression, and the dashed lines are 95% credible intervals for the predicted mean abundance of CWD at different tree densities.

Box 5.1
Simple linear regression for coarse woody debris

Coarse woody debris (CWD) is an important component of waterbodies, providing habitat for aquatic invertebrates and vertebrates. Christensen *et al.* (1996) examined the relationship between the abundance of CWD and tree density along the shoreline of 16 North American lakes. Quinn and Keough (2002) describe the analysis of these data using frequentist methods. Here I provide a Bayesian alternative. The model is very similar to the analysis of a mean (Box 3.2), with a simple modification to account for the linear relationship between the expected value of each lake (mean[i] in the code below) and the explanatory variable.

The mean is a function of two regression coefficients, each of which must be given priors. It is still assumed that the precision is the same for all the samples. Using uninformative priors for the three parameters that need to be estimated (the two regression coefficients and the precision), leads to the following code:

```
model
{
  a ~ dnorm (0, 1.0E-6) # prior for the intercept
  b ~ dnorm (0, 1.0E-6) # prior for the slope
  prec ~ dgamma(0.001, 0.001)     # prior for the
                                      precision
  for (i in 1:16)                  # for each of the
                                      16 lakes
  {
    mean[i] <- a + b*TreeDens[i]    # the mean CWD is a
                                      function of the
                                      tree density
    CWD[i] ~ dnorm(mean[i], prec) # amount of CDW,
                          drawn from a normal dist'n
  }
}
```

The data are given as:

```
list(TreeDens = c(1270, 1210, 1800, 1875, 1300, 2150,
1330, 964, 961, 1400, 1280, 976, 771, 833, 883, 956),
CWD = c(121, 41, 183, 130, 127, 134, 65, 52, 12, 46, 54,
97, 1, 4, 1, 4))
```

Initial values for the Markov chain (Chapter 1) were:

```
list(a=0, b=0, prec=100)
```

A total of 100 000 samples after discarding a burn-in of 10 000 samples led to the following output from WinBUGS:

Node	Mean	sd	2.5%	Median	97.5%
a	−77.04	32.97	−142.3	−77.13	−11.6
b	0.1154	0.02524	0.06528	0.1155	0.1652
prec	7.585E-4	2.869E-4	3.056E-4	7.218E-4	0.001418

Because uninformative priors were used, the results are numerically similar to those obtained by Quinn and Keough (2002).

This uncertainty needs to be accounted for when calculating the uncertainty associated with a prediction.

Fortunately in Bayesian statistics, the relatively complicated formula required for calculating the uncertainty of a prediction when using frequentist methods (Box 5.2) can be ignored. This is because by using WinBUGS, it is straightforward to calculate the uncertainty of a prediction. It requires only one more line of code to obtain the predicted mean CWD at a tree density of 1500, for example:

```
predicted_mean <- a + b*1500
```

Each sample of a and b from the Markov chain produces a different value of predicted_mean, so the uncertainty in the parameters a and b is propagated through to the predicted mean. The distribution of the parameters and any correlation between them (linear or otherwise) are taken into account. The variable predicted_mean can be sampled within WinBUGS just like any other. The 95% credible interval of CWD per km of shoreline is 73.1–119 m^2 when the tree density is 1500.

The above addition to the code predicts the mean amount of CWD at a given tree density, but does not incorporate the additional uncertainty associated with a new sample. Each sample is subject to additional variation around the mean, which can be included in the WinBUGS model by drawing each observation from a normal distribution with a specified precision. Therefore, if we wished to predict the tree density of a new

---

Box 5.2
Uncertainty of a prediction in frequentist analysis

In frequentist statistics, the uncertainty in the prediction for a particular value of the explanatory variable is measured by the standard error. This standard error of the prediction from the regression line ($s_{\hat{Y}}$) is estimated by using the formula:

$$s_{\hat{Y}} = \sqrt{s_{Y \cdot X}^2 \left[\frac{1}{n} + \frac{(X_i - \bar{X})^2}{\sum x^2} \right]},$$

where $s_{Y \cdot X}^2$ is the mean squared error of the residuals (the average squared deviation of the data from the regression line), n is the sample size, X_i the value of the explanatory variable for the prediction, $\bar{X}$ is the mean of the explanatory variable in the data used for the regression and $\sum x^2$ is the sum of squares of the explanatory variable. The equation for the standard error is more complicated, including the Gaussian multipliers (obtained by taking the inverse of the variance-covariance matrix), for regressions that contain more than one explanatory variable.

 The complexity of the formula is required because it is necessary to account for the uncertainty in the parameter estimates. It also accounts for the fact that it is easier to make precise estimates in the middle of the data rather than towards the edge of the domain. This complex formula can be ignored when using Bayesian methods.

---

sample, we could add one more line of code that draws the observation from the same distribution that is used in the model:

```
predicted_obs ~ dnorm(predicted_mean, prec)
```

WinBUGS takes the predicted mean that was calculated above and draws the prediction of the new observation from a normal distribution with that mean and the estimated precision. This precision measures the variation of each sample around the regression line.

 For the model considered in Box 5.1, the 95% credible interval for a new observation of CWD is 14.8–177 at a site with tree density of 1500 trees per km, which is considerably wider than the uncertainty around the mean. Thus, the variable predicted_mean estimates the uncertainty of the regression line (equivalent to a confidence interval for a regression) while the variable `predicted_obs` also accounts for the extra variation of observations around this (uncertain) regression line (equivalent to a prediction interval; see Sokal and Rohlf, 1995). This extra source of uncertainty inflates the credible interval.

As a final note on this example, readers might notice that the regression predicts negative abundances of CWD at low tree density. For example, the lower bound of the 95% credible interval is negative when tree density is less than approximately 900 (Fig 5.1). This is not ideal, but the example was used to compare the results of an equivalent frequentist analysis (Quinn and Keough, 2002). One possible interpretation is that any predictions of negative abundances correspond to predictions of no CWD debris. This is somewhat of an approximation that is not necessary when using Bayesian methods. It would be relatively easy to constrain the predicted mean to non-negative values and use a probability distribution that is also non-negative (e.g. a lognormal). Such regression models are described in more detail in the later part of this chapter.

Multiple linear regression

In simple linear regression, relationships are examined between one explanatory variable and a response variable. It goes without saying that more than a single variable might influence the response variable. For example, the amount of coarse woody debris in a lake might be influenced by characteristics of the lake (e.g. frequency of floods that might deposit coarse woody debris from inflowing streams).

Additional variables can be added to linear regressions by simply adding extra terms. Therefore, the effect of two explanatory variables (x_1 and x_2) on the response variable can be expressed as:

$$y = a + b_1 x_1 + b_2 x_2.$$

An important aspect of this model is that the effect of each variable on y is the same regardless of the value of the other variable. For example, the response variable y increases by b_2 units for an increase of one unit of the variable x_2, regardless of the value of the other parameter (x_1). Graphically, the above model implies that the relationship between y and x_2 is a series of parallel lines for different values of x_1 (Box 5.3).

Box 5.3
Multiple linear regression

Paruelo and Laueroth (1996) examined the relationship between the distribution of C3 grasses in the western portion of North America and several geographic variables. Quinn and Keough (2002) presented a multiple regression analysis of these data using longitude (LONG) and latitude (LAT) as explanatory variables, which will be repeated here. This analysis describes linear trends in the geographic distribution of the relative abundance of C3 grass species at 73 sites. WinBUGS code for this analysis is given below. Note that the explanatory variables have been 'centred' by subtracting the means. This is helpful in many WinBUGS analyses for improving the efficiency of the sampling algorithms and has no effect on the results (see Box 5.8). Further, the response variable (relative abundance of grass species) has been transformed prior to analysis to remove problems of extrapolating to negative relative abundance scores (see following sections on non-linear models):

```
model
{
  mLONG <- mean(LONG[])
  mLAT <- mean(LAT[])
  for (i in 1:73)
  {
    Y[i] <- log(C3[i] + 1)
    Y[i] ~ dnorm(mean[i], prec)
    mean[i] <- a + b[1]*(LONG[i]-mLONG) + b[2]*
            (LAT[i]-mLAT)
  }
  a ~ dnorm(0, 1.0E-6)
  for (i in 1:2)
  {
    b[i] ~ dnorm(0, 1.0E-6)
  }
  prec ~ dgamma(0.001, 0.001)
}
```

The initial values were:

```
list(a=0, b=c(0,0), prec=100)
```

After excluding the initial 10 000 as a burn-in (Box 1.4), 100 000 iterations leads to the following parameter estimates for a, b[1] and b[2].

Node	Mean	sd	95% CI
a	0.22	0.017	0.19–0.25
b[1] (longitude)	−0.0010	0.0028	−0.0065–0.0044
b[2] (latitude)	0.025	0.0033	0.019–0.032

There is strong evidence for a relationship between latitude and the relative abundance of grass species, with the relative abundance of C3 grasses increasing with distance from the equator. The predicted relationship with longitude is relatively weak; the most extreme effect based on the 95% credible interval is a decline of 0.0065 for each increase in degree longitude. Such a change results in a reduction of approximately 0.17 across the range of the data, which is considerably smaller than the predicted (mean) change with latitude (0.57).

The predicted weak relationship with longitude can be calculated (with credible intervals) for a given latitude by adding a few lines of code. The prediction below is based on latitude 35° North, although any value could be substituted. Predictions are made across a range of values for longitude at intervals of one degree. The code propagates uncertainty in the estimation of a, b[1] and b[2] through to the prediction:

```
for (i in 93:120)
{
  predlat35[i] <- a + b[1] * (i-mLONG) +
                  b[2] * (35-mLAT)
}
```

These predicted effects can be sampled in WinBUGS and plotted in a graphics package. For example, the predicted relationship between latitude and the relative abundance of C3 grasses is shown with 95% credible intervals in Fig. 5.2 for two different latitudes (35° and 45° North). The predicted relationships for the two different latitudes are parallel because the model only included additive effects. If the relationship between relative abundance and longitude differs at different latitudes, then an interaction term would be required and the lines would not be parallel (see Box 5.4).

If results were required in the original units (relative abundance) rather than the logarithmically transformed values, the values predlat35[i] could be back-transformed with the following line of code included within the above 'for loop':

```
predrichlat35[i] <- exp(predlat35[i]) - 1
```

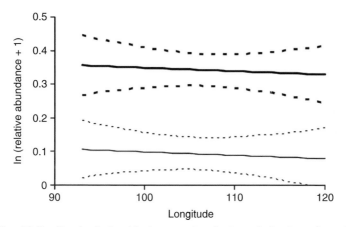

Fig. 5.2 Predicted relationship between longitude and the (transformed) relative abundance of C3 grasses in North America for different latitudes (thin lines, 35° North; and thick line, 45° North), ignoring interaction terms. The dotted lines are the 95% credible intervals of the regression lines.

Interaction terms

Instead of assuming a consistent effect of each variable on the relative abundance of C3 grasses, the effect of each variable could depend on the value of the other. This dependence is known as an interaction between the variables. The simplest formulation for an interaction is to assume that the regression coefficient for each variable is itself a linear function of the other variable. Thus, b_1 could be written as $c_1 + k_1 x_2$ and b_2 as $c_2 + k_2 x_1$. Substituting these expressions into the above regression equation leads to:

$$y = a + (c_1 + k_1 x_2)x_1 + (c_2 + k_2 x_1)x_2,$$

which gives:

$$y = a + c_1 x_1 + c_2 x_2 + (k_1 + k_2)x_1 x_2.$$

Therefore, linear interactions between two variables can be expressed by generating a new variable that is the product of the two. The regression coefficient for this new variable is equal to the sum of the regression coefficients k_1 and k_2 in the interactions. This implies that these two values (k_1 and k_2) cannot be estimated separately from each other, with the coefficient for the interaction reflecting their combined effect. The consequence of the interaction term is that the slope of the relationship between y and x_2 depends on the value of x_1 (and vice versa). Graphically, the series of lines (at different values of x_1) for the relationship between y and x_2 are not parallel if the interaction term is not equal to zero (Box 5.4).

If there were three explanatory variables, a second-order interaction term could be generated by multiplying the three variables together. A second-order interaction implies that the magnitude of one or more of the pair-wise (first-order) interaction terms depends linearly on a third variable. This means, for example, that the size of the effect of the first variable depends on a second variable, but the magnitude of this dependence is linearly related to a third variable.

It is sometimes difficult to interpret higher-order interaction terms. I prefer to select the interaction terms of interest a priori and then to ignore the others. The selection is based on the particular research questions that are being asked and whether the interaction terms are biologically meaningful. There is little point including an interaction term in a statistical model if there is no ecological basis for it or its effect cannot be meaningfully interpreted.

Box 5.4
Linear regression with an interaction term

For the model examined in Box 5.3, it was assumed that the change in relative abundance of C3 grasses with latitude was the same regardless of the longitude; the two predicted lines in Fig. 5.2 are parallel. This need not be the case. For example, the gradient in relative abundance of C3 grasses versus latitude may depend on whether the sites are close to the coast (higher longitudes) or inland. Such a situation would mean that the predicted effect of each variable (longitude and latitude) depends on the value of the other. If this dependence is linear, we can model the effect using a new variable that is equal to longitude×latitude. This can be included in the above WinBUGS code by changing how the mean is specified:

```
mean[i] <- a + b[1]*(LONG[i]-mLONG) + b[2]*(LAT[i]-
mLAT) + b[3]*(LONG[i]-mLONG)*(LAT[i]-mLAT)
```

This introduces an extra parameter ($b[3] = k_1 + k_2$) that requires a prior distribution. After taking 100 000 samples from the posterior distribution using this model, the parameter estimates for the regression coefficients are:

Node	Mean	sd	95% CI
a	0.216	0.017	0.183–0.250
b[1] (longitude)	−3.132E-4	0.0027	−0.0056–0.0049
b[2] (latitude)	0.029	0.0035	0.022–0.036
b[3] (interaction)	0.0014	5.387E-4	0.0003–0.0024

The DIC (see Chapter 4) for this model (-72.0) is less than that for the model without the interaction term (-67.5). The difference in DIC values is sufficiently large for us to believe that there is likely to be an interaction between latitude and longitude on their relationship with the relative abundance of C3 grasses.

The predicted relationship between relative abundance and longitude illustrates that the interaction is likely to be important, with a positive relationship at higher latitudes (e.g. 45° North) and a negative relationship at lower latitudes (e.g. 35° North). The interaction term means that the predicted effect of latitude is less in the central part of North America (e.g. 95° West) than in the western part (e.g. 110° West). This is illustrated by the greater difference between the two lines (35° and 45° North) at higher longitudes. Thus, the inclusion of the interaction term influences our understanding of how longitude is related to the relative abundance of C3 grasses. Without the interaction term, we would conclude that relative abundance does not vary substantially with longitude. However, by including the interaction term, we would conclude that the relative abundance of C3 grasses does appear to vary with longitude but the nature of that relationship depends on the latitude. At higher latitudes the relative abundance declines with distance from the coast, while at lower latitudes the relative abundance of C3 grasses increases with distance from the coast. (See Fig. 5.3.)

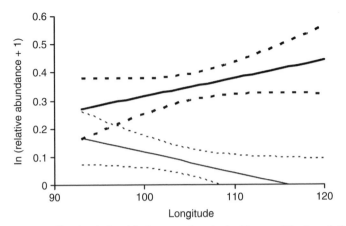

Fig. 5.3 Predicted relationship between longitude (degrees West) and the (transformed) relative abundance of C3 grasses in North America for different latitudes (thin lines, 35° North; and thick line, 45° North), including a linear interaction term between the two variables. The dotted lines are the 95% credible intervals.

Non-linear regression

It is not necessary to assume that there is a perfect linear relationship between the dependent variable and the explanatory variables. The simplest way to accommodate non-linearity is to transform the dependent variable to help make the relationship more linear. For example, one might conduct a regression of annual mortality of different raptor species versus body mass. However, if we assumed a linear relationship, then it is possible that extreme body masses would lead to predicted mortality rates greater than one or less than zero. Such results would be non-sensical.

In contrast, if we transformed mortality, then it might be possible to account for the required limits for a proportion. A common approach is to use a logit transformation for proportions, which converts them to a scale between minus infinity and plus infinity. The logit transformation is given by:

$$\text{logit}(p) = \ln(p/[1-p]).$$

This is also known as the log-odds ratio, being the logarithm of the odds ratio. The odds ratio is simply the probability of the event happening divided by the probability of it not happening.

It is important to bear in mind that probability values of zero or one are not defined under this transformation, because they lead to values of minus infinity and plus infinity respectively. Back-transformation gives the original probability p as a function of logit(p):

$$p = 1/(1 + \exp[-\text{logit}(p)]).$$

In Box 5.5, a multiple regression model of annual mortality illustrates how to transform the dependent variable in WinBUGS. Using the means of the posterior distributions for the regression coefficients leads to the predicted relationship between body mass and mortality for diurnal and nocturnal raptors (Fig. 5.4). For diurnal raptors, the predicted relationship for annual mortality is:

$$m = 1/(1 + \exp[0.217 + 0.9540 \times \text{Mass}]).$$

For owls it is:

$$m = 1/(1 + \exp[0.776 + 0.9540 \times \text{Mass}]).$$

In all the previous models, it was assumed that there was a linear relationship between the predicted values (or a transformation of the predicted values) and the explanatory variables. For example, models of the form:

$$y = a + bx + cx^2$$
$$y = a + \exp(-bx), \text{ or}$$
$$y = a + \sin(x),$$

have not been analysed.

It is relatively easy to analyse such models in WinBUGS. It is simply a matter of including the relevant function in the expression for the model. For example, it would be possible to express the relationship between tree density and coarse woody debris (Box 5.1) as a logistic function:

$$\text{expected CWD} = a/(1 + \exp(-(b + c(\text{TreeDens}[i] - m\text{Tree})))).$$

This equation is an s-shaped curve, with the parameter a being the maximum density of CWD, and b and c controlling the minimum CWD and the rate of change from the minimum to the maximum. An advantage of this model over the linear model is that the amount of CWD is constrained to be non-negative. Further, extrapolation at higher tree densities does not lead to unrealistically large abundances of CWD.

Box 5.5
Non-linear transformation of the dependent variable

Larger bodied animals tend to have lower mortality rates than
smaller animals (Peters, 1983; Calder, 1984; Savage *et al.*, 2004). The
relationship is not precise, but the information can help to determine
population parameters for many species, and can be particularly
helpful when data are scarce.

In this example, a regression of raptor mortality against body
mass is conducted. A second explanatory variable is also added to
account for possible differences in mortality rate between diurnal and
nocturnal raptors. Such a difference would be expected given that
persecution by humans is a common source of mortality for
raptors. By being active at night, owls might be less susceptible
to this persecution and have lower mortality rates. Flying and
foraging by night or day may lead to other differences that
might cause owls and diurnal raptors to have different mortality
rates.

Therefore, we have two explanatory variables to describe
variation in the annual mortality of different raptor species: body
mass and whether the species is an owl. The latter variable is coded as
one if the species is an owl and zero otherwise. Body mass is
measured in kilograms. Using the data on 26 species from McCarthy
et al. (1999), the following WinBUGS code can be used for the
analysis:

```
model
{
  for (i in 1:26) # for each of the 26 raptor species
  {
    lp[i] <- a + b*Mass[i] + c*Owl[i]    # linear
                                           predictor
    lm[i] <- logit(Mortality[i])         # logit
                      transformation of Mortality
    lm[i] ~ dnorm(lp[i], prec)           # assume logit
                                           (Mortality)
                                           is normally
                                           distributed
  }
  a ~ dnorm(0, 1.0E-6) # intercept term
  b ~ dnorm(0, 1.0E-6) # effect of body mass
```

```
    c ~ dnorm(0, 1.0E-6)           # effect of being an owl
    prec ~ dgamma(0.001, 0.001) # precision
}
```

This is very similar to a basic linear regression of two variables. The variable b defines the effect of body mass and c defines the difference in mortality between owls and diurnal raptors. The only difference is that the dependent variable (`Mortality[]`) is transformed using the logit function to become the new variable `lm[]`. It is this transformed variable that is assumed to be drawn from a normal distribution.

The data for the above model are:

```
list(Mass = c(0.37, 0.28, 0.64, 0.14, 1.2, 0.58, 0.9,
0.3, 1.6, 0.885, 0.53, 1.2, 0.3, 0.56, 0.7, 1.22, 0.94,
0.14, 0.22, 0.865, 0.95, 2.5, 2.3, 3.4, 1.04, 0.565),
Owl=c(1, 1, 1, 1, 1, 1, 1, 1, 0, 0, 0, 0, 0, 0, 0, 0, 0,
0, 0, 0, 0, 0, 0, 0, 0, 0),
Mortality=c(0.33, 0.15, 0.23, 0.3, 0.12, 0.2, 0.15,
0.31, 0.18, 0.3, 0.3, 0.25, 0.48, 0.39, 0.31, 0.22,
0.2, 0.48, 0.41, 0.25, 0.26, 0.04, 0.035, 0.08, 0.12,
0.25))
```

Initial values for the Markov chain were:

```
list(a=0, b=0, c=0, prec=100)
```

After discarding the first 10 000 samples as a burn-in, 100 000 samples provides the following estimates of the regression coefficients:

Node	Mean	sd	95% Bayesian CI
a	−0.2167	0.1742	−0.5606–0.128
b	−0.9543	0.124	−1.198−−0.709
c	−0.559	0.2047	−0.9653−−0.1543

The negative regression coefficients demonstrate that increasing body mass (b) and being an owl (c) both lead to lower annual mortality rates of raptors.

Note that using a transformation of the dependent variable is the only time that a node can be defined twice in WinBUGS. WinBUGS recognizes that a formulation of this type means that the transformed data are drawn from the specified distribution. The same results would have been achieved if a linear regression had been conducted using logit-transformed mortality (calculated outside of WinBUGS) as the dependent variable.

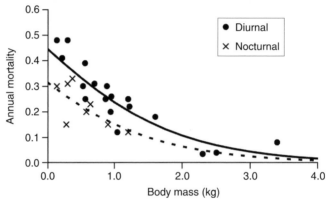

Fig. 5.4 Annual mortality of raptors versus body mass for diurnal (solid line) and nocturnal (broken line) raptors. The predicted lines are shown along with the original data (symbols) that were obtained from McCarthy *et al.* (1999).

The analysis of this model is described in Box 5.6. The model fits the data better compared with the linear model (Fig. 5.5), reducing the DIC value from 164.7 to 155.1. Thus, the non-linear model is a demonstrably better fit and it is more realistic.

In frequentist analysis, it is difficult to calculate confidence intervals for predictions of non-linear regression lines. Various approximations are required. By using Bayesian methods, the intervals around the regression can be calculated in the same manner as for the linear models by calculating predicted values across the range of the data.

Logistic regression

Analyses based on normal distributions have been used extensively in ecology, primarily because of their analytical convenience. However, in many circumstances, ecologists record the presence or absence of a species or an event (e.g. the occurrence of fire), and the parameter to be modelled is the probability of the event occurring. For example,

---

Box 5.6
Non-linear regression for coarse woody debris

The WinBUGS code for estimating the parameters of the non-linear
regression of coarse woody debris (CWD) on tree density is given
below. Uninformative priors are used for the parameters, with the
parameters a and c constrained to be non-negative. Abundance of
CWD is predicted for tree densities of between 800 and 2100 trees/km,
from which credible intervals around the regression line can be
calculated by monitoring the variable pred[] within WinBUGS. It is
assumed that the abundance of coarse woody debris is drawn from a
lognormal distribution to ensure that the values are not negative.

```
model
{
  a ~ dunif (0, 1000)        # uninformative prior,
                               constrained to be
                               positive
  b ~ dnorm (0, 1.0E-6)      # uninformative prior
  c ~ dunif(0, 10)           # uninformative prior,
                               constrained to be
                               positive
  prec ~ dgamma(0.001, 0.001) # uninformative prior
  mTree <- mean(TreeDens[])   # mean of the
                                explanatory variable
  for (i in 1:16) # for each lake sampled
  {
    pred[i] <- a/(1 + exp(-(b+c*(TreeDens[i]-
          mTree))))              # predicted CWD
    logpred[i] <- log(pred[i]) # take the logarithm
                                 of the prediction
    CWD[i] ~ dlnorm(logpred     # CWD drawn from
          [i], prec)            lognormal
  }
  for (i in 8:21) # make predictions fro tree
                    densities 800- 2100
  {
    pred[i] <- a/(1 + exp(-(b+c*(i*100-mTree))))
  }
}
```

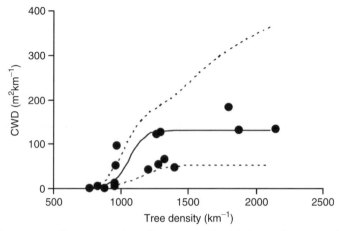

Fig. 5.5 Non-linear regression of coarse woody debris in lakes versus the tree density. The solid line is the prediction for the mean parameter estimates and the dashed lines are 95% credible intervals on the predictions. The wide credible interval at high tree densities occurs because the exponential transformation inflates larger predictions more than small predictions, and the relative paucity of data.

Parris (2001) examined the habitat of *Litoria pearsoniana*, a stream-breeding frog of eastern Australia. Of particular interest was how the probability of occurrence of the species varied with the size of the stream and the nature of the vegetation near the stream.

When the data are presences or absences (represented as ones and zeroes) and the prediction is the probability of occurrence, the deviations from the predicted regression line would not be even approximately normally distributed. This makes linear regression inappropriate, but logistic regression is suitable (McCullagh and Nelder, 1989). In logistic regression, the prediction is the probability of occurrence (or, more gene-rally, the probability that the event occurs). The name logistic regression arises because a logistic function is used to convert a linear combination of the explanatory variables to the probability of occurrence. This is the same function as used in the *Non-linear regression* section, but in logistic regression it is assumed that the data are binary outcomes with a given probability.

In other respects, logistic regression proceeds in the same way as when conducting linear regression. Additional terms such as non-linear functions, extra explanatory variables and interaction terms can be included in the model. An example of using logistic regression to model the occurrence of a rare plant is given in Box 5.7.

Box 5.7
Logistic regression: The occurrence of *Leionema ralstonii* in rock outcrops

In her Ph.D. thesis, Jane Elith used logistic regression to model the distribution of *Leionema ralstonii*, a plant species associated with steep rocky outcrops in forests of southeastern New South Wales, Australia (Elith, 2002). Data were obtained from 325 sites, with the species found at 73 of those. Two possible sets of explanatory variables were considered, with the best-fitting set having five explanatory variables. These variables were:

numrock   – the number of rocky outcrops within 400 m of the site
volcanic – a variable reflecting the mapped geology
rockcell – the number of GIS cells that were coded as outcrop within a radius of 100 m
nearslope – the maximum slope within a radius of 50 m of the site
nearsouth – a measure of the extent to which the site faces south.

The presence or absence of *Leionema ralstonii* was coded by using a one or zero. The variable nearslope was included as both linear and quadratic variables to account for a possible non-linear relationship. The WinBUGS code for this analysis is:

```
model
{
  a ~ dnorm(0, 1.0E-6)       # the intercept term
  for (i in 1:6)             # the 6 regression
                                 coefficients
  {
    b[i] ~ dnorm(0, 1.0E-6)  # uninformative priors
  }
  mnr <- mean(numrock[])     # means of the explanatory
                                 variables
  mv <- mean(volcanic[])
  mrc <- mean(rockcell[])
  msl <- mean(nearslope[])
  mso <- mean(nearsouth[])
```

```
for (i in 1:325)          # for each of the sites
{
  logit(p[i]) <- a + b[1]*(numrock[i]-mnr) +
  b[2]*(volcanic[i]-mv) + b[3]*(rockcell[i]-mrc)
  + b[4]*(nearslope[i]-msl) + b[5]*
  (nearslope[i]-msl)*
  (nearslope[i]-msl) + b[6]*(nearsouth[i]-mso)
  #logit(p) is a function of the explanatory
   variables and regression coeff's
    lr[i] ~ dbern(p[i]) # observed occurrence drawn
                          from a Bernoulli dist'n
}
}
```

In this example, the variables have been 'centred' by subtracting the mean (see Box 5.8). This helps to improve the efficiency of the MCMC algorithm by generating posterior samples with lower autocorrelation. The logit of the probability of occurrence of the species is a linear function of the explanatory variables. It is assumed that the observed occurrence of the species is determined randomly, using a Bernoulli distribution (dbern) that generates a one (presence) with the specified probability and a zero (absence) otherwise. The following parameter estimates were obtained from 100 000 iterations of the model:

Node	Mean	95% Bayesian CI
a	−3.18	−4.2−−2.3
b[1]	0.314	0.047–0.59
b[2]	3.30	2.1–4.8
b[3]	0.113	0.067–0.16
b[4]	0.0418	−0.0096–0.097
b[5]	0.00556	0.00054–0.011
b[6]	−0.0103	−0.019−−0.0016

The occurrence of *Leionema ralstonii* is positively associated with the variables numrock, volcanic and rockcell, and negatively associated with nearsouth. This reflects its association with north-facing volcanic outcrops.

The quadratic effect of nearslope produced a u-shaped relationship, with the occurrence minimized when the slope was approximately 20°. The highest rates of occurrence were on the

steepest slopes (up to approximately 50°). This reflects the species'
strong association with steeper areas, but also its occurrence in
riparian areas and along spurs where the ground is relatively flat. The
nature of the relationship between the occurrence of *Leionema* and
slope can be seen by plotting the predicted occurrence across the range
of the data (Fig 5.6). The 95% credible intervals on the predictions
illustrate the considerable uncertainty associated with the predictions
in flatter areas, and the high rates of occurrence in steeper areas.

Box 5.8
Centring data for better sampling

In some of the previous models, explanatory variables have been
centred by subtracting the mean. The advantage of this is that the
correlation between successive samples is reduced, which improves
the efficiency of the MCMC sampling. This can be illustrated by
using a simple logistic regression for the presence of lizards on islands
as a function of their perimeter-area ratio. The data were originally
analysed by Polis *et al.* (1998) and also presented in Quinn and
Keough (2002). The simplest possible model is given below.

```
model
{
  a ~ dnorm(0, 1.0E-6)  # uninformative intercept term
  b ~ dnorm(0, 1.0E-6)  # uninformative effect of PA
  for (i in 1:19)       # for each island
  {
    logit(p[i]) <- a + b*PA[i]  # logit(p) a function
                                  of PA
    Y[i] ~ dbern(p[i])            # observed occurrence
                                  drawn from
                                  Bernoulli dist'n

  }
}
```

The same model with the perimeter-area ratio being centred
would be:

```
model
{
  a ~ dnorm(0, 1.0E-6)  # uninformative intercept term
```

```
b ~ dnorm(0, 1.0E-6) # uninformative effect of PA
mPA <- mean(PA[]) # calculate mean perimeter:area
                        ratio
for (i in 1:19) # for each island
{
  # uses centred PA ratio logit(p) a function of
    (PA − mean)
  logit(p[i]) <- a + b*(PA[i] - mPA)
  Y[i] ~ dbern(p[i]) # observed occurrence drawn
                            from Bernoulli dist'n
}
  # re-calculates the original intercept term
  Intercept <- a - b*mPA
}
```

The regression coefficients for the slopes are unaltered by centring the data, but the intercept term is changed. The variable Intercept generates what would be the intercept term (a) in the original model.

Sampling from Markov chains can cause successive samples to be highly correlated. If there is positive correlation between the first and second samples in a chain and between the second and third, then there will also be positive correlation between the first and third, through their common association with the second sample. However, the correlation between the first and third will be less than the correlation between the successive samples. The 'lag' measures the number of samples between two samples in a Markov chain (e.g. successive samples have a lag of one). As the lag between samples increases, the correlation between samples will continue to decrease towards zero. A plot of the correlation versus the lag is known as an auto-correlation plot.

The benefit of centring the data can be seen in the auto-correlation plot for the variables in the model. For example, centring reduces the correlation to approximately zero for samples that are three or more iterations apart, but the correlation is still greater than zero for samples 20 apart when the data are not centred (Fig. 5.7). By centring the data, fewer samples are needed to obtain the same level of precision for the parameter estimates. This is because when the samples are correlated, the information about the posterior distribution contained in each new sample is similar to those values already sampled.

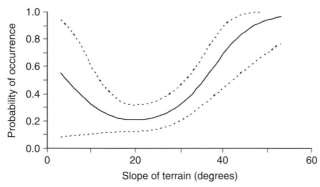

Fig 5.6 Probability of occurrence of *Leionema ralstonii* versus the slope of the terrain. The predicted relationship is shown for a volcanic substrate and while holding the other variables at their mean values. The solid line is the mean of the posterior predictions and the broken lines represent the 95% credible interval.

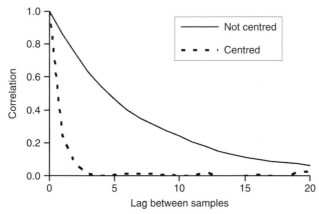

Fig. 5.7 Correlation between successive samples of the regression coefficient for the presence of lizards on islands versus the lag between samples. The autocorrelation plot is shown for models with and without centring, illustrating that centring reduces the correlation close to zero for samples that are three apart. The correlation is still greater than zero for samples 20 apart when the data are not centred.

Imperfect detection

In the example of logistic regression (Box 5.7), it was assumed that the species was recorded without error. This is reasonable given that the species is relatively distinctive. It is unlikely to have been confused with other species and its size means that it is unlikely to have been overlooked in the surveys. However, the same cannot be said for other species, where it is possible that a species may inhabit a site but remain undetected during a single survey.

The possibility of imperfect detection of a species raises the issue of false negative errors (Wintle *et al.*, 2004). Such errors might occur if a species uses the area being surveyed but is not present at the time of the survey or if it is present but has sufficiently cryptic characteristics that it remains undetected. In these circumstances the observer would record that the species was absent when it does in fact use the site.

By conducting multiple surveys of the same sites, it is possible to estimate both the probability of presence and the probability of detection conditional on the species being present (Mackenzie *et al.*, 2002; Tyre *et al.*, 2003; Wintle *et al.*, 2004, 2005a). This can be modelled as a two-stage process. First, the actual presence or absence of the species is modelled at each site. This may be conducted using the usual approach to logistic regression. Secondly, the number of times that the species is recorded at each site is modelled. If the species is present, we can assume that the species is detected on each visit with probability d. In such cases, it can be assumed the number of survey visits on which the species is recorded is drawn from a binomial distribution with parameters d (the probability of detection per visit) and v (the number of survey visits). If the species is absent, then the probability of detection would equal zero. The WinBUGS code for such an analysis is given in Box 5.9.

There are several advantages of accounting for imperfect detection in the analysis. The first is that not accounting for it would cause the

---

Box 5.9
Accounting for imperfect detection

Kirsten Parris studied the presence of *Litoria pearsoniana*, a tree frog that breeds in streams on the east coast of sub-tropical Australia. Multiple surveys were conducted at each of 64 sites, using two different survey methods. These two methods were nocturnal searches and the use of automated tape recorders from which the frogs could be identified by the advertisement calls of the males. The data required for each site are the number of surveys and the number of detections of the species with each method. The presence of the species was modelled as a function of the size of the stream (measured by the logarithm of the average annual volume of rainfall in the catchment upstream of the site, LnCV below) and the presence or absence of palms at the site as an indicator of mesic or xeric conditions. The WinBUGS code for an analysis of these data is given below:

```
model
{
  a ~ dnorm(0, 1.0E-6) # uninformative priors for the
                            variables
  b[1] ~ dnorm(0, 1.0E-6)
  b[2] ~ dnorm(0, 1.0E-6)
  b[3] ~ dnorm(0, 1.0E-6)
  d[1] ~ dunif(0, 1) # detection probabilities when
                          the species is present
  d[2] ~ dunif(0, 1)
  mLnCV <- mean(LnCV[]) # average catchment volume
  for (i in 1:64) # for each of the 64 sites
  {
    logit(p[i]) <- a + b[1]*(LnCV[i] - mLnCV) +
    b[2]*palms[i] + b[3]*(LnCV[i] - mLnCV)*palms[i]
    # probability of presence
    Lp[i] ~ dbern(p[i])  # actual presence
    dd1[i] <- d[1]*Lp[i] # detectability of nocturnal
                              searches
    dd2[i] <- d[2]*Lp[i] # detectability of automatic
                              tape recorders
    Y1[i] ~ dbin(dd1[i], V1[i])        # number of
                                        detections
                                        with searches
    Y2[i] ~ dbin(dd2[i], V2[i])        # number of
                                        detections
                                        with tapes
  }
}
```

In the above code, the variable b[1] represents the effect of stream size on presence of the frog, b[2] represents the effect of the presence of palms and b[3] is the coefficient for the interaction term so that the effect of stream size depends on whether palms are present.

The presence of the species is determined randomly and recorded as the variable Lp[i] for each site; a value of one indicates that the species is present while it is absent if the variable is zero. The variables d[1] and d[2] are the detection probabilities of the nocturnal searches and the tape recorder surveys if the species is present at the site, and dd1 and dd2 are the detection probabilities that depend on whether the species is present or absent at the site. The data Y1 and Y2 are the number of surveys during which the species was recorded by the two survey methods, and V1 and V2 are number of surveys at each site using the two methods (the identifier 1 refers to nocturnal searches, and 2 refers to tape recorders).

The parameter estimates obtained from 50 000 samples are given below, indicating the positive association with palms (b[2]) and the positive effect of stream size when palms are present (b[3]). The detection rate of nocturnal searches (mean = 0.561) is greater than that of automatic tape recorders (mean = 0.353).

Node	Mean	95% CI
a	−1.54	−2.9–−0.37
b[1]	0.588	−0.88–2.2
b[2]	2.44	0.89–4.3
b[3]	3.18	0.30–7.0
d[1]	0.561	0.47–0.65
d[2]	0.353	0.27–0.44

The predicted probability of occurrence across the range of stream sizes can be generated for when palms are present and also when they are absent. The code for this is given below, which may be inserted into the above model. When palms are present, the variables for the effect of palms (b[2]) and the interaction term (b[3]) are included. The predicted relationship is shown in Fig. 5.8.

```
for (i in 1:20)
{
  LVol[i] <- 2 + 3*i/20 # covers the range of stream
                          sizes
  logit(predpalms[i])    <- a + (b[1] + b[3])*
  (LVol[i] - mLnCV) + b[2]
  logit(prednopalms[i])  <- a + b[1]*(LVol[i] -
  mLnCV)
}
```

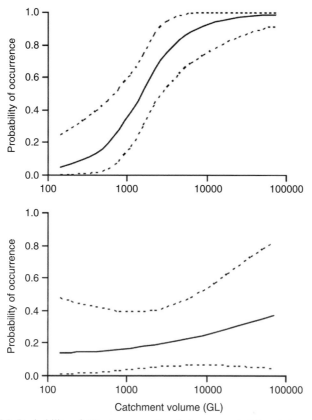

Fig. 5.8 Probability of *Litoria pearsoniana* being present at forest streams in eastern Australia as a function of stream size when palms are present (a) or absent (b). The analysis accounted for the imperfect detection of frogs on each survey, the two types of surveys and the different number of surveys at sites (based on data from Parris, 2001).

estimated occurrence to be underestimated. Secondly, if the probability of detection is not considered, it is necessary to assume that survey effort at all of the sites is identical. For example, in the study of *Litoria pearsoniana*, some sites were surveyed only four times, while others were surveyed up to 18 times. To use the data without accounting for effort would require, for example, only using the data from the most recent four surveys. This would mean deleting relevant hard-won field data, which is never a very satisfactory way to conduct analyses. Finally, different surveys methods may have different chances of detecting species. Analyses that consider the probability of detecting a species can accommodate these differences (see Box 5.9).

Poisson regression

In the previous chapter, the Poisson distribution was suggested as a useful model for describing counts of organisms or species. This distribution can also be used when analysing relationships between counts and explanatory variables. Therefore, rather than assuming that the response variable is normally distributed, as in linear regression, it would be possible to assume that the response variable is distributed as a Poisson random variable. The latter form of analysis is referred to as Poisson regression.

One feature of the Poisson distribution is that its parameter (which is equal to the mean and variance) is required to be positive. To achieve this in the regression, the usual procedure is to assume that the natural logarithm of the parameter (m) is equal to a linear combination of the explanatory variables (e.g. x_1 and x_2):

$$\ln(m) = a + b_1 x_1 + b_2 x_2$$

This may be transformed to:

$$m = \exp(a + b_1 x_1 + b_2 x_2),$$

which ensures that the value for m is positive.

For example, Ellison (2004) used Poisson regression to analyse ant species richness as a function of elevation, latitude, and habitat type (bog versus forest). The analysis is described in Box 5.10.

Box 5.10
Poisson regression: ant species richness

Ellison described a Poisson regression of ant species richness. The explanatory variables were latitude, elevation, and habitat type. In the WinBUGS code below, the continuous variables latitude and elevation have been centred to improve the efficiency of sampling. Without centring, there is strong autocorrelation in the samples, which means that more samples are required to achieve the same precision in the estimates. In fact without centring, the correlation is so strong that it leads to incorrect inference about the parameters unless an extremely large number of samples are taken (Ellison, 2004). The regression coefficients for the explanatory variables are unaltered by the centring, but the intercept term (`alpha`) is modified. However, the intercept term for the equivalent model without centring can be calculated (see code below).

Ellison (2004) used results from previous research (Brühl *et al.*, 1999; Gotelli and Arnett, 2000) to generate informative priors for the effect of latitude and elevation on ant species richness. The priors for the coefficients of these variables were represented as informative normal distributions. The explanatory variable 'habitat' was treated as a categorical variable with a value of one representing forest habitats and a zero representing bogs.

```
model
{
# uses centred data
  ml <- mean(lat[]) # calculates averages for
                            centring
  me <- mean(elev[])
  for(i in 1: N) # for each of the N data points
  {
    richness[i] ~ dpois(mu[i]) # ant species richness
                            drawn from a Poisson
    log(mu[i]) <- alpha + beta[1]*(lat[i]-ml) +
    beta[2]*(elev[i]-me) + beta[3]*habitat[i]
                # a log-linear model for the average
  }
  intercept <- alpha - beta[1]*ml - beta[2]*me
  # recovers the intercept term
  # informative priors
  preclat <- 1/(0.04 * 0.04) # precision of lat
                            effect, based on
                            s.e. of 0.04
  precelev <- 1/0.0003/0.0003 # precision of elev
                            effect
  alpha ~ dnorm(0.0,1.0E-6)    # uninformative
                            intercept
  beta[1] ~ dnorm(-0.1725,preclat) # informative
                            effect of
                            latitude
  beta[2] ~ dnorm(-0.0022,precelev) # informative
                            effect of
                            elevation
```

```
beta[3] ~ dnorm(0.0,1.0E-6) # uninformative effect
                                 of habitat
```

The posterior distribution of the parameter estimates had the following values based on 100,000 samples:

Node	Mean	sd	Bayesian 95% CI
alpha	1.507	0.09765	1.311–1.695
beta[1]	−0.1848	0.03322	−0.25−−0.1196
beta[2]	−0.001811	2.358E-4	−0.002273−−0.0013
beta[3]	0.6365	0.1197	0.4036−0.8732

Credible intervals for the informative priors, the likelihood function of the data, and the posterior distributions for the parameters beta[1] and beta[2] are shown in Fig. 5.9. Without the informative prior distribution, the results are less precise.

There is a suggestion that the prior for the effect of elevation is not consistent with the data because their 95% credible intervals only overlap slightly (Fig. 5.9, see also Belia *et al.*, 2005). A model using uninformative prior distributions for both parameters did not fit particularly better (DIC = 209.0 using uninformative priors versus 209.7 when using the informative priors).

Correlation

In regression, models are built to describe relationships among variables. However, in some cases, the question is simply what is the strength of the relationship between variables? In these circumstances, correlation analysis is useful. The most commonly used index of correlation is Pearson's product moment correlation coefficient, which will be considered here. This coefficient measures the strength of the linear relationship between two variables.

When there is a perfect positive linear relationship, the correlation coefficient equals 1, with a perfect negative linear relationship indicated by a value of −1. Intermediate values represent less than perfect linear relationships, with a value of zero representing no correlation. It is important to note that a zero correlation does not necessarily mean that there is no relationship between the variables, just that there is not a

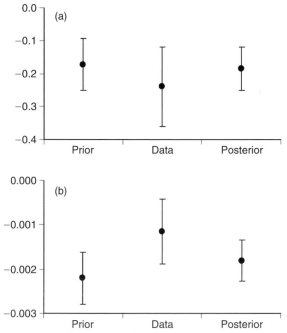

Fig. 5.9 The 95% credible intervals for the parameter estimates of the model of Ellison (2004) describing the effect of latitude (a) and elevation (b) on ant species richness. Intervals are shown for the prior, the likelihood (data) and the posterior. In both case, there is a negative effect, and using the prior has increased the precision of the effect.

linear component to any relationship. For example, a perfect symmetrical u-shaped relationship between two variables would generate a correlation coefficient of zero even though there is a well-defined relationship between the variables.

Pearson's product moment correlation coefficient is equal to the covariance of the two variables divided by their standard deviations. Details of how to calculate the covariance can be found elsewhere (e.g. Sokal and Rohlf, 1995), but these details are not necessary for analysing correlations in WinBUGS. In WinBUGS, the simplest analysis of the correlation coefficient is to assume that the two variables are drawn from a bi-variate normal distribution. Both variables have a mean and standard deviation that needs to be estimated. The correlation coefficient measures the strength of the relationship between the variables.

In WinBUGS, a bi-variate normal distribution is defined by the means of the two variables and the inverse of the variance-covariance matrix of the variables (see Box 5.11). This inverse matrix is the bi-variate equivalent of the precision used in the normal distribution.

WinBUGS code for conducting a correlation analysis is given in Box 5.11. It is relatively complex given that it is a simple correlation, but has a considerable advantage over the usual frequentist procedures

---

Box 5.11
Correlation analysis

Analysing correlations within WinBUGS is most easily achieved using a bi-variate normal distribution, so it may be necessary to transform the data to improve conformity to this distribution. Code for conducting the analysis is given below, in which an uninformative uniform prior is used for the correlation coefficient, wide positive uniform priors are used for the standard deviation of the variables, and uninformative normal distributions are used for the means.

```
model
{
  mean[1] ~ dnorm(0.0, 1.0E-6)
  mean[2] ~ dnorm(0.0, 1.0E-6)
  sd[1] ~ dunif(0, 100)
  sd[2] ~ dunif(0, 100)
  correl ~ dunif(-1, 1)
  covar[1,1] <- sd[1]*sd[1]
  covar[2,2] <- sd[2]*sd[2]
  covar[1,2] <- correl*sd[1]*sd[2]
  covar[2,1] <- correl*sd[1]*sd[2]
  omega[1: 2, 1: 2] <- inverse(covar[,])
  for (i in 1:N)
  {
    Y[i, 1:2] ~ dmnorm(mean[], omega[,])
  }
}
```

The bi-variate distribution is defined in WinBUGS by the means of the two variables and the inverse of the variance-covariance matrix (omega[ ]). The elements of the variance-covariance matrix are constructed from the standard deviations and the correlation coefficient. The diagonal elements of the matrix (covar[1,1] and covar[2,2]) are simply the variances of the two variables, and the two off-diagonal elements are both equal to the covariance.

Data on the relationship between crab biomass and the number of crab burrows at two sites on Christmas Island (Green, 1997; see also Quinn and Keough, 2002) are given below:

```
list(N=8, Y=structure(.Data=c(2.15, 39, 2.27, 38,
4.31, 61, 2.58, 79, 3.23, 35, 1.83, 39, 1.54, 45,
2.00, 28),.Dim=c(8,2)))
list(N=10, Y=structure(.Data=c(4.36, 38, 4.01,
37, 3.33, 27, 2.63, 18, 4.46, 41, 3.96, 33, 4.18, 40,
4.21, 29, 2.54, 25, 4.29, 38),.Dim=c(10,2)))
```

The 95% Bayesian credible intervals for the correlation coefficient between these two variables, based on 100 000 samples, were [–0.45, 0.81] and [0.42, 0.96] for the two sites, with the means of posterior distributions equal to 0.278 and 0.796. The considerable overlap in the credible intervals of the correlation coefficients suggests that they may be similar at both sites despite the relatively large difference between the means of their posterior distributions.

by generating precise estimates of the distribution of possible values. A wide range of frequentist methods has been developed to place confidence intervals on correlation coefficients and conduct hypothesis tests (e.g. Sokal and Rohlf, 1995). These methods are based on transformations and large-sample approximations. The null hypothesis required for the usual tests is that the correlation coefficient is equal to zero. Formulae are available for non-zero null hypotheses, with the approximation used depending on sample size (Sokal and Rohlf, 1995). In comparison to the number and complexity of the frequentist approximations available for analysing correlations, the analysis in WinBUGS is simple (Box 5.11) and does not need to be modified to account for small sample sizes or non-zero null hypotheses.

Model-based priors for correlations

It is possible in some circumstances to have prior information for correlation coefficients. For example, McCarthy (1997) examined the correlation between median natal dispersal distances predicted by models and those observed in the field for song sparrows and banner-tailed kangaroo rats. Two models were examined, one in which the density of vacancies influenced dispersal distances (exponential model) and one in which both vacancies and the density of competitors (simultaneous model) influenced dispersal distances (McCarthy, 1997).

McCarthy (1997) relied on traditional null hypothesis testing to compare the models. There was a positive correlation between the predictions and observations for the competition model for three different datasets and negative correlations for the exponential model. Statistically significant correlations were not obtained, but the results were not entirely satisfactory because the power of the tests was low. In fact, simulations conducted by McCarthy suggested that given the expected variation, large correlations were unlikely to be observed even if the models were a perfect description of reality.

It would be possible to use the models to generate informative priors and evaluate these priors with Bayesian methods. The construction of the informative priors for the correlation accounts for the fact that a perfect correlation between predictions and observations is very unlikely because of sampling error in the observed median dispersal distance of the individuals (Box 5.12).

Box 5.12
Model-based priors for correlations

The following code can be used to generate the expected correlation between the observed median dispersal distances and the predicted values if the model were a perfect description of reality. The model simulates the observed median distance for each year by randomly generating dispersal distances for each individual and then finding the median of the distances within each year. Both these steps are relatively complex, so I will describe them in some detail.

The first step in simulating the median dispersal distances is achieved by using the inverse of the cumulative distribution function (see Appendix B). In the code below, the exponential model is illustrated. For this model, the cumulative distribution function is given by $F(x) = 1 - e^{-vx}$, where v is the density of vacancies and x is the dispersal distance (see McCarthy, 1997). Re-arranging this expression leads to the inverse of the cumulative distribution function $x = -\ln[1 - F(x)]/v$. This expression allows us to randomly generate

---

dispersal distances, by substituting a uniform random number (between zero and one) for the term $(1–F(x))$. This is one of the standard methods for generating a random number when the inverse of the cumulative distribution function can be obtained. (see also Press *et al.*, 1992; Knuth, 1997).

The second step takes the randomly generated dispersal distances and calculates the median for each year. WinBUGS does not have a function for calculating the median, so a method is given below. When the sample size (n) is even, the median is the average of the $n/2$th and ($n/2+1$)th measurement when they are ranked from the smallest to the largest. When the sample size is odd, the median is equal to the (($n+1)/2$)th measurement. For example, if $n = 9$, the median is the fifth measurement when they are ranked.

The following code uses these two steps to calculate the expected correlation between the observed medians and predictions if the model were a perfect description of reality. The predicted median dispersal distance in each year (`predmedian[]`) is calculated externally using the formula provided by McCarthy (1997) for that model. For example, the median dispersal distance is predicted to equal $\ln(2)/v$ for the exponential model, where v is the density of vacancies.

```
model
{
  for (i in 1:nyears)
  {
    for (j in 1:N[i])
    {
      # random number used in the following generation
        of dispersal distances
      rand[i, j] ~ dunif(0, 1)
      # dispersal distance for exponential model
      D[i,j] <- -log(rand[i,j])/v[i]
    }
    # The following calculates the median simulated
      dispersal distance in each year
    m0[i] <- trunc(N[i]/2)
    m1[i] <- m0[i]+1
    # use the following for the median if N[i] is odd
    median1[i] <- ranked(D[i, 1:N[i]], m1[i])
```

```
# use the following for the median if N[i] is even
median0[i]  <-  (ranked(D[i,  1:N[i]],  m0[i])  +
median1[i])/2
# the expression ''even[i]'' is equal to 1 if N[i]
  is even, and 0 if odd
even[i] <- equals(m0[i], N[i]/2)
simmedian[i] <- even[i]*median0[i] +
(1-even[i])*median1[1]
}
# now calculate correlation between predicted and
  observed median dispersal distance
meanpred <- mean(predmedian[])
meansim <- mean(simmedian[])
for (i in 1:nyears)
{
  y1[i] <- predmedian[i] - meanpred
  y2[i] <- simmedian[i] - meansim
}
covarsim <- inprod(y1[], y2[])/(nyears-1)
correl <- covarsim/sd(predmedian[])/
sd(simmedian[])
}
```

For example, for male song sparrows and the exponential model, the correlation between predictions and observations is unlikely to be greater than approximately 0.5 and may in fact be negative even when the model is a perfect description of reality (Fig. 5.10). The data used for this analysis are:

```
list(nyears=10,
predmedian=c(1.155,  1.333,  1.575,  1.000,  1.998,
2.082, 2.476, 0.788, 0.963, 1.333),
v=c(0.6,  0.52,  0.44,  0.693,  0.347,  0.333,  0.28,
0.88, 0.72, 0.52),
N=c(19, 17, 24, 29, 18, 19, 15, 18, 21, 19)
```

For the simultaneous dispersal model, the dispersal distance depends on the density of vacancies (as for the exponential model) but also the density of dispersers. The code for this more complicated model is provided on the website, but follows the same pattern as that above.

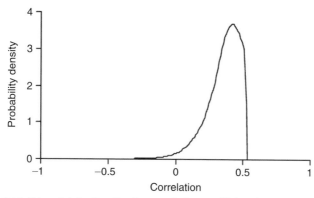

Fig. 5.10 Prior distribution for the correlation coefficient between observations and predictions of the exponential model for male song sparrows.

The DIC values for the analyses based on the informative priors can be compared to the DIC values for an analysis in which the correlation coefficient is equal to zero and an analysis in which an uninformative prior is used. The results of these analyses help to indicate whether the given model of dispersal is a good predictor of natal dispersal distance, compared with a model that predicts no relationship, or some relationship that is different from that predicted.

The correlation between the predictions of the simultaneous model and observed median dispersal distance was examined for three datasets (male and female song sparrows and banner-tailed kangaroo rats; see McCarthy, 1997). The prior with the lowest DIC value varied between datasets, with each of the three priors (uninformative, zero, and informative) being best for one of the datasets (Table 5.1). However, the DIC values are additive across the three analyses, so it is possible to obtain an overall measure of the predictive performance of the priors. In this case, the uninformative prior had the lowest total DIC, with the informative prior close behind (DIC was 0.68 units greater), and an analysis with zero correlation being worst. Thus, there was some support for the simultaneous model because the correlation between its predictions and observations appeared to be different from zero. However, the predictions of the simultaneous model were no better than a model that predicted a non-zero (uninformative) correlation, so it is possible that an alternative (and as yet unformulated model) might make better predictions.

For the exponential model, the correlations between its predictions and the observed data were best described by analyses in which the

Table 5.1. *DIC values obtained when examining correlations between the observed median dispersal distance and the predictions of the simultaneous dispersal model and the exponential dispersal model. The results are shown for the three data sets examined and for three different priors (uninformative, zero and an informative prior based on the dispersal model).*

Prior	Male song sparrow	Female song sparrow	Kangaroo rat	Total	ΔDIC
Simultaneous model					
uninformative	115.76	129.10	50.10	294.96	0
zero	115.06	129.00	54.18	298.24	3.28
informative prior	115.44	128.54	51.66	295.65	0.68
Exponential model					
uninformative	127.61	145.94	67.31	340.86	0.65
zero	126.9	145.62	67.69	340.21	0.00
informative prior	128.28	146.47	69.71	344.46	4.25

correlation was zero, and the exponential dispersal model was consistently poor (the total difference of DIC values was 4.65), suggesting that it was not a particularly good model of natal dispersal distances.

Thus, the analyses suggest that the simultaneous dispersal model is a reasonable model, with the correlation between its predictions and observations being noticeably different from zero. However, the exponential model appeared to be a poor model, with the correlation between its predictions and observations being best described as zero. Therefore, the Bayesian analysis lends support to the simultaneous dispersal model, especially relative to the exponential dispersal model.

Concluding remarks

Regression analyses in this chapter build on the analyses of means and frequencies, by using functions to describe how the predicted values change in response to explanatory variables. Linear regression results when this function is assumed to be linear and deviations around the regression line are assumed to be normally distributed. Other forms of regression arise when non-linear relationships and other probability distributions are used. Logistic, Poisson, and non-linear regression were described in this chapter.

Correlation analysis is also used to study relationships among variables. Rather than using one or more explanatory variables to model the response of a dependent variable, correlation analysis simply describes association. The most common form of analysis is to assume that the variables being examined are normally distributed, with the correlation coefficient measuring the strength of the linear association between them. Regression analyses examine how variables measured on a continuous scale influence the expected value of another variable. In the next chapter, the influence of categorical variables is examined with analysis of variance (ANOVA).

6

Analysis of variance

In regression analysis, the explanatory variables are usually continuous variables. However, in some of the previous analyses, a categorical variable with two possible values was included. For example, in the analysis of the ant data (Box 5.10), two different habitat types were considered by using the variable 'habitat', which takes a value of zero for forests and one for bogs. Similarly, the habitat model for the frog *Litoria pearsoniana* used a variable that indicated whether palms were present or not. These types of explanatory variables are referred to as factors. The analysis of a variable in response to one or more factors is known as analysis of variance or ANOVA.

It is not my intention to cover all aspects of ANOVA here, but to introduce the most basic elements and then to discuss how some of the more advanced aspects can be included with ease within WinBUGS. More details about using ANOVA in ecology can be found elsewhere (Sokal & Rohlf, 1995; Fowler *et al.*, 1998; Underwood, 1997; Quinn and Keough, 2002). Readers unfamiliar with ANOVA are advised to read these or similar books. In particular, I will not cover much of the theory behind the use of fixed or random factors, repeated measures and interaction terms, but focus on how they can be analysed with Bayesian methods.

One-way ANOVA

Imagine we are interested in analysing how the masses of starlings vary between four different locations (Fowler *et al.*, 1998). To study this, an ecologist would weigh different starlings from each location. In the case of Fowler *et al.* (1998), ten starlings were measured from each. A model for analysing such data might be that the mass of starlings differs between the locations. The model would specify the mean mass at each location and assume that variation in mass between different birds was described by a particular probability distribution. If a normal distribution was used, an additional assumption (to help reduce the number of estimated parameters) might be that the variances of the distributions for the different habitat were all identical.

A model of this form, with one factor describing differences, is known as a one-way or one-factor ANOVA. The explanatory variable would be an index to represent the different locations, and the response variable would be the mass of the starlings. The model could be coded easily in WinBUGS (Box 6.1).

Coding of variables

In Box 6.1, the mean for each location was specified directly. This formulation of the problem worked perfectly well for this simple analysis. However, other formulations are useful for some analyses, especially when extra factors are added (Quinn and Keough, 2002). Useful formulations include specifying how the mean at each location differs from a global mean or how the mean at each location differs from a reference location. The reference location might be chosen arbitrarily (e.g. the first or last location) or depending on the comparisons of interest (e.g. urban locations might be compared to locations in natural areas such as grasslands, heathlands, and forests, making the urban location a useful choice for reference). The actual choice does not influence the results, but may make interpretation more straightforward.

The analysis of the starling data conducted in Box 6.1 is repeated in Box 6.2 by using the fourth location as a reference class. In this case, a variable is added to represent the mean of the reference location, and differences between each location and this reference location are estimated. The difference for the fourth location (i.e. the reference location) is set to zero. The change does not make a substantive difference to the results, with the same means and DIC values being generated. The main difference is that values being estimated ($d[1]$, $d[2]$, and $d[3]$) now represent how the mean mass of the first three locations differ from the fourth, rather than the actual masses of each location.

Box 6.1
A simple one-way ANOVA

The code below can be used to conduct a one-way ANOVA for data
on the mass of 40 starlings sampled from four different locations. The
variables a[1], a[2], a[3], and a[4] are the means of the masses
at the four different locations.

```
model
{ # simple one-way ANOVA
  for (i in 1:4)              # the mean for each of
                               the four locations,
                               drawn from
                               uninformative
                               priors

  {
    a[i] ~ dnorm(0.0, 1.0E-6)
  }
  tau ~ dgamma(0.001, 0.001)  # uninformative
                               precision
  for (i in 1:40)             # for each of the
                               40 birds sampled

  {
    mean[i] <- a[location[i]]  # the mean depends on
                               the location
    Y[i] ~ dnorm(mean[i], tau) # mass drawn from
                               a normal
                               distribution

  }
}
```

The above code has a different mean mass for each of the four
locations. The indicator variable location has values of 1, 2, 3 or 4
in the data (data on website), depending on where each of the 40 birds
was sampled.

For the data provided by Fowler *et al.* (1998), the DIC
value for this model is 221.2, which is considerably smaller than a
model in which the means did not differ among the locations
(DIC = 237.3). Therefore, there is strong evidence to suggest
that the mass of the starlings differs between the four
locations. The mean mass at each of the four locations can

be obtained from the variables a[1] – a[4], and the means and standard deviations of their posterior distributions are given below:

Node	Mean (g)	sd (g)
a[1]	83.6	1.16
a[2]	79.4	1.16
a[3]	78.6	1.16
a[4]	75.4	1.15

Box 6.2
Simple ANOVA using a reference class

The model developed in Box 6.1 is reproduced here, but using the fourth location as the reference class.

```
model
{ # one-way ANOVA, using reference class
  base ~ dnorm(0, 1.0E-6) # the mean of the reference
                            class
  for (i in 1:3)          # the mean for each of the
                            first 3 locations,
                            relative to the
                            reference class
  {
    d[i] ~ dnorm(0.0, 1.0E-6)
  }
  d[4] <- 0 # no diff b/t the mean of the 4th location
             and the reference class
# because the fourth location is the reference class
  tau  ~  dgamma(0.001,  0.001)  #  uninformative
                                    precision
  for (i in 1:40) # for each of the 40 bids sampled
  {
    mean[i] <- base + d[location[i]] # the mean
                               depends on the
                               location
```

```
    Y[i] ~ dnorm(mean[i], tau)
  }
}
```

DIC values generated from the above code and that in Box 6.1 are identical. The difference in the results is that the values of d[1], d[2] and d[3] now represent how the mean mass at the first three locations differ from that at the fourth location (the reference class), which is given by the variable base. After taking 100 000 samples from the posterior distribution, the mean mass at the first three locations appears to be greater than at the fourth.

Node	Mean	sd
d[1]	8.2	1.63
d[2]	4.0	1.63
d[3]	3.2	1.63
base	75.4	1.15

The actual means of each of the different locations can be recovered by incorporating extra variables. For example, the mean mass of starlings at the first location would be equal to base+d[1]. If these values were recorded, the posteriors for the mean masses of the four locations would be identical to the values obtained in Box 6.1. Readers may wish to confirm this for themselves, although it is unnecessary to do this for the mean mass at the fourth location, which equals base+d[4]. Because d[4] in this analysis is set to zero, the variable base is equal to the mean mass at the fourth location; it has the same posterior distribution as a[4] from Box 6.1.

Fixed and random factors

In analysing relationships between variables, there are two types of explanatory variables that need to be considered. These are known as fixed factors and random factors. Fixed factors are those where all types of treatments are included in the analysis. If the study were to be conducted again, the same treatments or types would be included in the analysis. Random factors are different because only a selection of the types or treatments is included in the analysis, and a different selection of these could be made if the study was replicated. The distinction between the two types of variables is important because it influences what can be inferred from the analysis.

The difference between the two types of variables can be illustrated by the starling example. The topic of interest was the variation in the mass of starlings between different locations. Four different locations were chosen for the analysis. How these were chosen depends on whether the variable location should be treated as a fixed or random factor. If the study was conducted because we are particularly interested in these four locations and no others, then the variable location would be included in the analysis as a fixed factor. For example, if the four locations were classified as forest, heathland, grassland and urban, and we were interested in comparing the mass of starlings at these particular types of locations, then the variable location would be a fixed factor. If the study was repeated, starlings would again be sampled from these four types of locations.

However, if the analysis was motivated by the expectation that all locations might differ, perhaps regardless of their classification, then the chosen four would be simply a random (or perhaps haphazard) sample from all possible locations. In this case, any set of four locations would be adequate for the study and the variable location would be treated as a random factor. If the study were replicated, a set of different locations would be chosen.

Random factors are easily analysed in WinBUGS. The simplest approach is to assume that the mean of each location (in the starling example) is drawn from a distribution, with a particular mean and standard deviation. Then, within each location, each sample is drawn from a distribution with a mean given by its location and a particular standard deviation. For analytical convenience, it is usually assumed that the standard deviation of the samples within each location is the same for all locations, although this is not a necessary assumption when using Bayesian methods.

WinBUGS code for analysing the starling data with location as a random factor is given in Box 6.3. Two standard deviations need to be estimated, one which measures the variation in mass of starlings among locations and one that measures the variation between starlings within locations.

Note that this is another example of a hierarchical model (Box 3.6), with the global mean and standard deviation among locations being hyper-parameters.

Box 6.3
One-way ANOVA with a random factor

The simplest approach to modelling a random factor is to
specify how the mean deviates from a global mean for each
value of the factor. For example, consider the starling data
from Box 6.1. In this case, we would include a global mean
mass, and then specify how the mean of each location
differs from this value. Then, as a second step, we specify
how the actual mass of each starling deviates from the mean of
each location.

Normal distributions are usually used when modelling
random effects, although other distributions could also be used.
In the case of the starling data, two different standard deviations
are required. The first specifies how the mean mass of starlings
varies among locations and the second how the mass varies
between individuals within locations. The code for this analysis
is given below.

```
model
{# random effects one-way ANOVA
  a ~ dnorm(0, 1.0E-6) # a is the global mean mass
  for (i in 1:4) # the deviation from the global mean
                  of the mean mass at each location
  # drawn from normal with estimated level of
    variation
  {
    d[i] ~ dnorm(0.0, tau_a)
  }
  sd_among ~ dunif(0, 100) # uninformative sd -
                              variation among
                              locations
  tau_a <- 1/(sd_among*sd_among) # convert sd to
                                    precision
  sd_within ~ dunif(0, 100) # uninformative sd -
                              variation between
                              starlings within
                              locations
  tau_w <- 1/(sd_within*sd_within) # convert sd to
                                      precision
```

```
for (i in 1:40) # for each of the 40 birds sampled
{
  mean[i] <- a + d[location[i]] # the mean
                                    depends on the
                                    location
  Y[i] ~ dnorm(mean[i], tau_w) # data drawn from
                                    normal dist'n

}
}
```

The variation among and within locations can be investigated by analysing the two standard deviations. The means of the posterior distributions and the 95% credible intervals for these standard deviations are given below.

Node	Mean	95% CI
sd_among	6.46	1.71–22.7
sd_within	3.69	2.93–4.73

It is clear that there is more uncertainty about the level of variation among locations, with the 95% credible interval for the standard deviation among locations more than encompassing that of the standard deviation within locations. The greater uncertainty arises because there are only four locations that can be used to estimate the variation among locations, while there are 40 starlings to estimate the variation within locations.

Two-way ANOVA

ANOVA can be extended to include more than one factor. For example, data on the mass of starlings could be sampled at the same four locations a second time. The total variation in the data can then be described by differences among time periods and differences among locations. The simplest model in this case is where there are additive effects of the two factors. Code for analysing this two-factor analysis is given in Box 6.4.

Box 6.4
Two-way ANOVA

In the case where the starling data are obtained from an additional time period, a second variable is needed, which I will call 'period'. For multi-way ANOVAs, it tends to be easiest to use a reference class to develop the model. Otherwise, it is difficult to ensure that the correct number of estimated parameters is included. Code for analysing the two-way ANOVA for the starling data, using both location and period as fixed factors, is given below.

```
model
{# two-way ANOVA, using reference classes
  base ~ dnorm(0, 1.0E-6)
  for (i in 1:3) # the mean for each of the four
  locations, expressed as a difference from the
  reference class
  {
    a[i] ~ dnorm(0.0, 1.0E-6)
  }
  a[4] <- 0
# the mean for each of the two time periods, expressed
  as a difference from the reference class
# the first time period is used as the reference class
  b[1] <- 0
  b[2] ~ dnorm(0.0, 1.0E-6)
  tau ~ dgamma(0.001, 0.001) # uninformative
                                   precision
  for (i in 1:80) # for each of the 80 birds sampled
  {
    mean[i] <- base + a[location[i]] + b[period[i]]
            # the mean depends on the location and
            period
    Y[i] ~ dnorm(mean[i], tau)
  }
}
```

Samples from the posterior distribution of the variables a (for location) and b (for period) indicate that the first three locations and second period have higher body masses.

Node	Mean	95% credible interval
a[1]	7.4	4.8–9.9
a[2]	5.0	2.4–7.6
a[3]	3.6	1.0–6.2
b[2]	9.1	7.3–10.9
base	75.3	73.2–77.3

The DIC for this model was 458.7.

Interaction terms in ANOVA

The two-way model considered in the previous section included only additive effects. This means that the effect of each factor does not depend on the value of the other factor. To relax this assumption, it is possible to include an interaction term that allows the effect of each factor to depend on the value of the other factor. This is equivalent to the use of interaction terms in regression models.

When including interaction terms, it is important to ensure that the correct number of parameters is included. Only a certain number of parameters can be uniquely estimated. A more complete discussion of this topic is provided by Quinn and Keough (2002). The important point is that when using reference classes in ANOVA, interaction terms are not estimated for those cases that include the reference class. This ensures that the model is not over-parameterized (see Quinn and Keough, 2002 for further details). The analysis of a two-way ANOVA with interaction terms is illustrated in Box 6.5.

Variance partitioning

In the case of the random factor model for the body mass of starlings, the variation was separated into variation among locations and variation within locations. This is known as variance partitioning. Examining the sources of variation is useful, for example, to help decide where effort should be placed to obtain an improved estimate. If there is greater variation in body mass among locations than within locations, then to obtain a more precise estimate of the global mean body mass of starlings,

Box 6.5
Including an interaction term in ANOVA

The model in Box 6.4 included only additive effects. It was assumed that the effect of each location was same in both time periods. However, it is possible that the effect of each location depends on the time period (or vice versa). A model that includes this possibility can be analysed by using interaction terms.

When using interaction terms, the means for the different combinations of classes are completely independent of each other. When using reference classes to define the model, the easiest way to achieve this is to add an extra term for each combination of classes. However, to ensure that the correct number of parameters is estimated, it is necessary to set the interaction term to zero for those cases that involve one or more of the reference classes. In the starling example, the reference classes were chosen as the fourth location and the first time period. This choice of reference classes is arbitrary, and the results would be identical if different reference classes were chosen. The code for the two-way ANOVA with interaction terms is given below.

```
model
{# two-way ANOVA with interaction term, using
    reference classes
  base ~ dnorm(0, 1.0E-6)
  for (i in 1:3) # the mean for each of the four
                      locations, expressed as a
                      difference from the
                      reference class
  {
    a[i] ~ dnorm(0.0, 1.0E-6)
  }
  a[4] <- 0
# the mean for each of the two time periods, expressed
  as a difference from the reference class
# the first time period is used as the reference class
  b[1] <- 0
  b[2] ~ dnorm(0.0, 1.0E-6)
# interaction terms - these are set to zero for cases
  involving one or more reference classes
  for (i in 1:3)
```

```
{
  int[i,1] <- 0
  int[i, 2] ~ dnorm(0.0, 1.0E-6)
}
int[4,1] <- 0
int[4,2] <- 0
tau ~ dgamma(0.001, 0.001) # uninformative
                             precision
for (i in 1:80) # for each of the 80 birds sampled
{
  mean[i] <- base + a[location[i]] + b[period[i]] +
  int[location[i], period[i]] # the mean depends on
                               the location and
                               period
  Y[i] ~ dnorm(mean[i], tau)
}
}
```

The parameter estimates for this model, using the starling data, are given in the following table.

Node	Mean	95% CI
a[1]	8.2	4.5–11.9
a[2]	4.0	0.4–7.6
a[3]	3.2	−0.4–6.9
b[2]	8.8	5.1–12.4
base	75.4	72.8–78.0
int[1,2]	−1.61	−6.77–3.55
int[2,2]	1.99	−3.14–7.13
int[3,2]	0.78	−4.41–5.93

The credible intervals for the interaction terms encompass zero, suggesting that their addition to the model may not have made a substantial improvement. This suggestion is supported by examining the DIC value, with this model having a value that is 4.1 units greater (462.8) than the model without the interaction terms. Therefore, the additive model (Box 6.4) would be chosen as the most parsimonious model. There is little suggestion that the effect of location on the body mass of starlings differs between the two sampling periods.

it may be important to obtain data from more locations rather than more data from each location. The relative effort that should be expended depends on the cost of establishing a new location, the cost of measuring the mass of each starling and the level of variation within and among locations. Underwood (1997) demonstrates how to determine the number of locations and samples per location so that the design is optimally efficient for a given budget.

Variance partitioning can also help when deriving prior distributions from existing data. Consider the case where we wish to estimate the annual survival of a bird from its body mass. We may have survival estimates for several bird species, with perhaps more than one estimate for each species. There are several sources of variation that need to be considered in this case. One source of variation is the variation among species that can be explained by differences in body mass. This is analogous to the variation among raptor species that was described in Box 5.4.

The variation around the regression line arises from three sources. Firstly, it is unlikely that the average survival rate of each species will fall exactly on the regression line. There are several other factors that mean that a particular species will have a survival rate that is above or below average for its body mass. Secondly, each study of a given species is unique. There will be certain local and temporal factors (e.g. food availability, predator abundance, weather, and biases of the particular researcher or method) that mean that the survival rate of the birds in each study will differ even if the same species were being studied. Thirdly, every study has a finite sample size so there is at least some imprecision for each individual study. The observed survival rate will differ from the true survival rate of the birds being studied. These three sources of variation around the regression line need to be considered explicitly if the data are used to obtain a realistic prediction of the survival rate of a species (Box 6.6).

An example of ANOVA: effects of vegetation removal on a marsupial

In this section, I provide a detailed account of an ANOVA that is used to investigate the effect of removing vegetation on the capture rate of mulgara, a small marsupial of arid Australia.[1] The study is based on an

[1] This example is drawn from a paper by McCarthy and Masters (2005).

Box 6.6
Partitioning variation in the data

Data on survival rates of European passerines were used to predict the annual survival of the European dipper based only on its body mass. The prediction was made using a regression of survival versus body mass for 27 species of passerine from 47 studies. Data were obtained from the appendix of Johnston *et al.* (1997) for annual survival and Dunning (1993) for body masses. Only survival estimates in Johnston *et al.* (1997) that included standard errors were included in the analysis because the reliability of the others could not be ascertained. The estimate of annual survival of European dippers provided in Johnston *et al.* (1997) was also excluded.

Multiple studies of annual survival were available for some species, which were included by treating studies as a random factor nested within each species. Similarly, the species effect was treated as a random factor. Therefore, the total variance around the regression line of annual survival on body weight included variation due to species, variation among studies, and variation within studies. This latter source of variation was estimated by the standard error of the estimate provided in Johnston *et al.* (1997).

Normal distributions were assumed for the random effects and body mass was log transformed to improve linearity of the relationship. The annual survival of European dippers for a new study was predicted for a bird of the appropriate body mass (59.8 g) and accounting for the fact that the species and study effects were random. The WinBUGS code for this analysis is given below.

```
model
{
  sdst ~ dunif(0,10)        # variation among studies
                              within species
  sdspp ~ dunif(0,10)       # variation among species
  taust <- 1/(sdst * sdst)  # precisions
  tauspp <- 1/(sdspp * sdspp)
  a ~ dnorm(0, 1.0E-6)      # intercept term
  b ~ dnorm(0, 1.0E-6)      # effect of body mass on
                              survival
```

```
for (i in 2:28) # for each of the 27 species
                    (numbered 2 to 28)
{
  spp[i] ~ dnorm(0, tauspp) # random species effect
}
for (i in 1:47) # for each survival estimate
{
  study[i] ~ dnorm(0, taust) # random study effect
  p[i] <- a + b*log(weight[i]) + spp[species[i]] +
  study[i] # predicted survival based on body
            weight and random effects
  taub[i] <- 1/(se[i]*se[i]) # within study
                              variation estimated
                              from stated s.e.
  surv[i] ~ dnorm(p[i], taub[i]) # survival assumed
                              to be normally
                              distributed

}
# prediction for European dipper, based on body mass
  of 59.8 g
dipspp ~ dnorm(0, tauspp)
dipstudy ~ dnorm(0, taust)
dippred <- a + b*log(59.8) + dipspp + dipstudy
}
```

The predicted annual survival rate for a new study (dippred, ignoring the within study variation) had a mean of 0.57 and a standard deviation of 0.073. Note that many samples in WinBUGS may be required before they converge on the posterior distribution, especially if the initial values are not close to the centre of the posterior distribution.

In Chapter 7, the predicted annual survival rate of European dippers is used as an informative prior for an analysis of mark-recapture data. As will be shown, the prediction based on body mass substantially improves the precision of the estimate obtained from the mark-recapture study.

experimental manipulation conducted by Dr Pip Masters (Masters *et al.*, 2003) and also draws on prior information derived from an earlier observational study.

Understanding how removal of vegetation affects mulgara is important because the managers of a resort in arid Australia were harvesting hard spinifex (*Triodia basedowii*, a common species of grass) and using the clippings as mulch on garden beds. There was some concern that the harvesting of spinifex might reduce abundances of a number of native species, so an experimental study was undertaken to investigate this possibility.

Compared to some other scientific disciplines, experiments in ecology are relatively rare, especially those based in the field. This is because they are often expensive, involve logistical hurdles and tend to produce data with considerable variation. In such cases, it would be helpful if useful prior information could contribute to interpretation of experimental results. This is particularly so because prior information is often a motivating factor when conducting experiments. For example, the experiment conducted by Masters *et al.* (2003) was partly motivated by a previous observational study in which mulgara (*Dasycercus cristicauda*) were captured more frequently on sites that had last burnt 11 years previously than on sites that had burnt within the last year (Masters, 1993). The main structural difference in the vegetation between the two types of sites was a marked reduction in cover, particularly that of hard spinifex. Masters (1993) had found that higher spinifex cover was associated with higher capture rates of mulgara.

The previous observational data suggested that the capture rates of mulgara in the recently burnt area were on average approximately one quarter of those in the longer unburnt area. Therefore, the experimental removal of spinifex was expected to cause a similar reduction. Using a frequentist ANOVA, Masters *et al.* (2003) did not detect a statistically significant reduction in the capture rate of mulgara in the experimental study, although the data were suggestive of an effect ($P = 0.15$) and the observed difference appeared to be similar to that in the observational study. How might a Bayesian approach to this analysis help to clarify the available evidence?

A Bayesian analysis requires the specification of models that may explain the data that were observed in the experiment. One model that could be chosen is the same as the null hypothesis used by Masters *et al.* (2003), that being of no effect of the experimental manipulation on the capture rate of mulgara (model A).

A second group of models is that the experiment causes a change in the capture rate of mulgara. There are at least two possible ways of specifying an effect of the experiment. The first of these is that the effect of the experiment on mulgara is the same as that recorded in the observational study (model B). The second model with an effect is that we have no prior information that would allow us to predict the magnitude or even direction of the experimental effect (model C). Model C is equivalent to the alternative hypothesis used by Masters *et al.* (2003) in their null hypothesis test.

It is somewhat pessimistic to say we have no prior information given the observational data suggest that there is an effect of spinifex removal. However, such a point of view could be justified. For example, we might be uncertain about the possible confounding role of pseudo-replication in the observational study (as with any unplanned fire, there was spatial structure in the arrangement of sites; Masters, 1993). Alternatively, the observed reduction may have been caused by some other effect of the fire instead of the reduction in spinifex cover.

These three models (A, no effect; B, an effect consistent with the observational study; and C, some effect that cannot be predicted a priori) can be considered as three competing points of view. By using the results of the experiment, we hope to be able to help discriminate between them to determine which model or models are best supported by the data.

Repeated measures ANOVAs were used to analyse the experimental data, with the effect of the treatment being expressed in the interaction terms for time and treatment (Masters *et al.*, 2003). Essentially, this means that an effect of the treatment would be detected if the difference between sites in the two treatments differed between time periods (particularly between the pre- and post-treatment time periods). There were multiple observations at each site, and the variable site was treated as a random factor (see Box 6.7 for the WinBUGS code) to account for the fact that the capture rate might vary randomly from site to site due to differences other than their spinifex cover.

To reflect a lack of prior information in model C, an uninformative prior for the effect of spinifex removal was specified by using a normal distribution with a mean of zero and large standard deviation (1000) for the parameter that described the effect. The large standard deviation means that the parameter can take any value (i.e. positive or negative effects of the experiment were permitted), with essentially no prior influence on the actual value.

Box 6.7

An example of repeated-measures ANOVA

The following code was used to analyse the experimental habitat manipulation of Masters *et al.* (2003). The capture rate of mulgara was $\ln(x+1)$ transformed to help satisfy assumptions of normality. Mulgara were surveyed in one time period before the experimental manipulation and twice afterwards. In the absence of an experimental effect, changes in capture rates from the first to the subsequent time periods would not depend on whether the plots were treated or kept as controls. Therefore, any effect of the treatment will be expressed in the interaction between time and treatment, which are given by the variables r1 and r2 in the following code.

```
model
{
# Priors
  y ~ dnorm(0, 1.0E-6) # global reference capture rate
# the mean difference between treatment and control
  plots prior to the manipulation
  r0 ~ dnorm(0, 1.0E-6)
# r1 and r2 are interactions terms for time*treament
# chose either zero (no effect), uninformative
  priors, or informative priors
# r1 <- 0 # r1 = 0 means no effect of harvesting in 1st
  post-harvest sampling period
# r2 <- 0 # r2 = 0 means no effect of harvesting in 2nd
  post-harvest sampling period
# r1 ~ dnorm(0, 1.0E-6) # uninformative priors
# r2 ~ dnorm(0, 1.0E-6)
  r1 ~ dnorm(-0.74, 19.389) # informative priors from
                          Masters 1993 - using se
for prior (~0.227)
  r2 ~ dnorm(-0.74, 19.389)
  t1 ~ dnorm(0, 1.0E-6) # fixed time effect for first
                          period after treatment
  t2 ~ dnorm(0, 1.0E-6) # fixed time effect for second
                          period after treatment
  tau ~ dgamma(0.001, 0.001) # precision of residual
  sdsite ~ dunif(0, 10)
```

```
    tausite <- 1/sdsite/sdsite # precision of random
                                 site effect
  for (i in 1:10) # for each plot
  {
    re[i] ~ dnorm(0, tausite) # random plot (i.e.,
                                 site) effect
# calculate mean expectation for each plot in each of
  the 3 periods
# post-treatment means are calculated as differences
  from pre-treament means
# accounts for treatment, site and time effects
# t1 and t2 are the time effects
# r1 and r2 are the post treatment effects
# r1 and r2 are equivalent to the time*treatment
  interaction in Masters et al. (2003)
# Treat[] is 0 for control sites and 1 for treatment
  sites
    mean0[i] <- r0*Treat[i] + y + re[i] # the mean
capture rate in each plot prior to the treatment
    mean1[i] <- mean0[i] + r1*Treat[i] + t1 # mean
capture rate in each plot in the first period after
the treatment
    mean2[i] <- mean0[i] + r2*Treat[i] + t2 # mean
capture rate in each plot in the second period after
the treatment
# Observed data, ln(x+1) transformed
    Before[i] ~ dnorm(mean0[i], tau)
    After1[i] ~ dnorm(mean1[i], tau)
    After2[i] ~ dnorm(mean2[i], tau)
  }
}
```

For the model with an effect consistent with the observational study (model B), the prior distribution was taken from the posterior distribution of the effect of spinifex removal in the observational study. This was obtained by analysing the observational data of Masters (1993) in the same manner as the experiment (i.e. the data on capture rate were $\ln(x+1)$ transformed before conducting a repeated measures ANOVA that included the effect of sites as a random factor). The difference in the log-transformed capture rate between the sites with different spinifex cover in the observational study (i.e. between those sites recently burnt and those long unburnt) was estimated (see Box 6.8 for the WinBUGS code). The mean difference was a reduction of 0.74 in the more recently burnt area, with the standard deviation of the estimate being 0.23. A normal distribution with this mean and standard deviation was used

---

Box 6.8
An ANOVA to establish an informative prior

The informative prior for the effect of spinifex removal in the mulgara study was obtained from an observational study where recently burnt sites had lower spinifex cover than long-unburnt sites. Each of the six sites was surveyed once in each of 12 time periods. The model accounts for random differences among sites and differences in capture rates among time periods. The effect of lower spinifex cover (recent burning) is represented in the WinBUGS code below by the variable r. The posterior distribution for this parameter was used as an informative prior for the effect of spinifex removal in the experimental study (Box 6.7).

```
model
{
# Priors - all uninformative
  sdsite ~ dunif(0, 10) # std dev of site to site
                            variation
  a ~ dnorm(0, 1.0E-6) # intercept term
  r ~ dnorm(0, 1.0E-6) # effect of recent burn
  tau ~ dgamma(0.001, 0.001) # precision of residuals
  tausite <- 1/(sdsite * sdsite)
  for (i in 1:6) # for each site
  {
    SiteEffect[i] ~ dnorm(0, tausite) # random site
                                        effects
  }
  for (i in 2:12) # for each time period
  {
    TimeEffect[i] ~ dnorm(0, 1.0E-6) # fixed time
                                        effects
  }
  TimeEffect[1] <- 0
```

```
for (i in 1:72) # for each of the 12 time periods and 6
sites
  {
  # calculate the expected recapture rate (ln (x+1))
    transformed for each observation
  # expectation accounts for differences in fire
    history, time period and site
  m[i] <- a + r*RecentBurn[i] + TimeEffect[Time[i]]
  + SiteEffect[Site[i]]
  # Y[]'s are the observed ln(x+1) transformed
    capture rates
  Y[i] ~ dnorm(m[i], tau)
  }
}
```

Table 6.1. *Deviance information criteria (DIC) for the three competing models used to describe the observed data in the experimental study of the effects of habitat manipulation on mulgara conducted by Masters et al. (2003).*

Model	DIC
No effect	62.8
Effect consistent with observational study	58.0
Uncertain effect	61.8

as the informative prior in model B. For model A (no effect of the experiment), the parameters that specified the reduction in the capture rate of mulgara were set to zero.

The model in which the observed effect of the experiment was consistent with the observational data (model B) had the lowest DIC value (Table 6.1), suggesting that this model is the best explanation of the data. The next best model was one of an uncertain effect (model C), and the difference in the DIC value ($\sim$3.8) suggested that it had less support than model B. The difference in the DIC values for the two inferior models was approximately 1, suggesting that they are largely indistinguishable. If the prior information were ignored, we would be unable to discriminate reliably between the hypothesis of no effect (model A) and the alternative of an effect (model B), with only weak evidence in favour of the latter.

This is similar to the conclusion of Masters *et al.* (2003) where the *p*-value was 0.15. However, by using the observational data and finding that the experimental results were consistent with those data, the ability of the experiment to determine whether there was an effect was increased. By considering both the observational and experimental studies, we can be reasonably sure that removal of spinifex reduces the capture rate of mulgara.

The estimated effect of the removal of spinifex cover, taking into account the prior information from the observational study, was that the capture rate of mulgara is reduced by approximately 0.75 in both time periods when using $\ln(x+1)$ transformed data (0.78 in the first and 0.74 in the second). With back transformed results (Box 6.9), the posterior

---

Box 6.9

Back-transforming the predictions of a model

The response variable used for the analysis of the mulgara habitat manipulation was $\ln(x + 1)$ transformed. The mean capture rate of mulgara in each of the three time periods and in both control and treatment plots can be estimated by inserting the following code into the WinBUGS model given in Box 6.7.

```
meanC0 <- exp(y) - 1 # mean in control plots before
                               treatment
meanC1 <- exp(y + t1) - 1 # mean in control plots after
                               (t=1)
meanC2 <- exp(y + t2) - 1 # mean in control plots after
                               (t=2)
meanT0 <- exp(r0 + y) - 1 # mean in treatment plots
                               before treatment
meanT1 <- exp(r0 + y + r1 + t1) - 1 # mean in treatment
                               plots after (t=1)
meanT2 <- exp(r0 + y + r2 + t2) - 1 # mean in treatment
                               plots after (t=2)
```

The variable y is used for the capture rate in the reference class (control plots prior to manipulation), t1 and t2 and the effects of the two time periods after manipulation, r0 is the mean difference between control and treatment plots prior to manipulation, and r1 and r2 are the effects of the manipulation in the treatment plots.

distribution implies that the removal of spinifex reduced the capture rate of mulgara to approximately one-quarter of the original rate (Fig. 6.1). The standard deviation of the posterior distribution for the effect in both time periods was 0.20. This standard deviation is only slightly less than that of the prior (0.23), which indicates that the prior had a large influence on the estimated effect of vegetation removal in the experiment. Therefore, conclusions about the effects of spinifex removal have only been clarified to a small extent by the experiment. The observational data provide the most compelling evidence, with the experimental data being consistent with this prior evidence. This approach encourages precautionary management and helps to avoid the misinterpretation that *p*-values greater than 0.05 provide evidence in favour of the null hypothesis (no effect).

Analysis of covariance

So far we have encountered two types of explanatory variables, continuous and categorical. When continuous variables are analysed with regression-based methods, the analysis is often referred to as simply 'regression analysis'. When categorical variables are used, the analysis is referred to as 'analysis of variance' (ANOVA). When both categorical and continuous variables are used together, it is referred to as 'analysis of

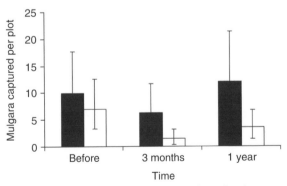

Fig. 6.1 Posterior distributions of the mean number of mulgara captures in treatment (open columns) and control (filled columns) plots, using the informative prior. Samples were taken before the treatment, and both three months and one year after the treatment. The columns give the means of the posterior distributions and the bars are the 95% credible intervals.

covariance' (ANCOVA). However, all these methods have the same basic structure and are jointly referred to as linear models.

Examples of ANCOVA have already been provided. In the case of habitat modelling for *Litoria pearsoniana*, stream size (continuous) and the presence of palms (categorical) were both included (Box 5.9). For analysing the survival of European passerines, study and species (both categorical) and body mass (continuous) were included (Box 6.6). In this section, ANCOVA is explored in more detail.

In the regression of raptor mortality (Box 5.5), owls and diurnal raptors both shared the same regression coefficient that described the effect of body mass on mortality. However, it is easy to imagine that the relationship between mortality and body mass might differ for the two different groups of raptors. ANCOVA accounts for this possibility by allowing the regression relationship to vary among two or more types of individuals.

An analysis of covariance of the owl mortality data can be conducted by simply substituting a choice of two possible regression coefficients, depending on whether the species is an owl or not (Box 6.10).

Box 6.10
Analysis of covariance

The code for the owl mortality regression (Box 5.5) is easily modified to allow the effect of body mass on the mortality to differ between owls and diurnal raptors.

```
model
{
  for (i in 1:26) # for each of the 26 raptor species
  {
    lp[i] <- a + b[Owl[i]+1]*Mass[i] + c*Owl[i]
          # linear predictor
    lm[i] <- logit(Mortality[i]) # logit
                                    transformation
                                    of Mortality
    lm[i] ~ dnorm(lp[i], prec) # assume
                                logit(Mortality) is
                                norm. dist'd
  }
```

```
a ~ dnorm(0, 1.0E-6) # intercept term
b[1] ~ dnorm(0, 1.0E-6) # effect of body mass for
                          diurnal raptors
b[2] ~ dnorm(0, 1.0E-6) # effect of body mass for
                          owls
c ~ dnorm(0, 1.0E-6) # effect of being an owl
prec ~ dgamma(0.001, 0.001) # precision
}
```

There are now two regression coefficients for the effect of body mass, with b[2] being used for owls and b[1] being used for diurnal raptors. The posterior distributions for these parameter estimates are very similar, which can be seen in the following summary that was obtained from 100 000 samples after excluding the first 10 000 as a burn-in.

Node	Mean	sd	95% CI
a	-0.22	0.18	-0.58–0.14
b[1]	-0.95	0.13	-1.2–0.69
b[2]	-1.01	0.49	-1.98–0.03
c	-0.52	0.37	-1.25–0.20

Given the similarity of the regression coefficients (b[1] $\approx$ b[2]), it is not surprising that the model with a common effect of body mass (Box 5.5) has a lower DIC value (35.6) than the ANCOVA model that requires an extra parameter (37.9).

ANCOVA: a case study

The first experiment

Forrester and Steele (2004) studied the effect of competition and resource availability on the mortality of gobies. The relationship between goby mortality and population density was analysed at three levels of resource availability, which in the case of gobies is the density of refuges that allow them to hide from potential predators. It is expected that mortality will increase as population density increases because of intraspecific competition for refuges. Because each refuge is only used by a single goby, it is expected that the effect of increasing the population density would be lower when the availability of refuges is high.

Forrester and Steele (2004) analysed their data using ANCOVA and assumed that the data were drawn from a normal distribution. As discussed previously, such an assumption is not well justified for proportional data, but the same assumption will be used for the Bayesian re-analysis for the sake of comparison (Box 6.11).

Box 6.11
ANCOVA example for goby mortality

Forrester and Steele (2004) conducted an experiment in which they manipulated the density of gobies and examined the effect of con-specific density on the mortality rate at three different densities of refuges. The original ANCOVA model analysed by Forrester and Steele (2004) allowed the intercept term and the slope to vary depending on the refuge density. This model has seven parameters. Given the relatively small sample size for the analysis (18 measurements of goby mortality), simplified models in which the intercept terms are identical are analysed here instead. This constraint (that the mortality is the same in the absence of competitors regardless of refuge density) is reasonable because refuges are unlikely to be limiting when there are no competitors. Two models were considered; one in which the effect of con-specifics was the same for all three refuge densities and one in which the effect differed. This latter model has five estimated parameters, while the former has three. The more complex model can be analysed using the code.

```
model
{
  a ~ dnorm(0, 1.0E-6)      # intercept
  b[1] ~ dnorm(0, 1.0E-6)   # effect of density at
                              resource level 1 (low)
  b[2] ~ dnorm(0, 1.0E-6)   # effect of density at
                              resource level 2
                              (medium)
  b[3] ~ dnorm(0, 1.0E-6)   # effect of density at
                              resource level 3 (high)
  prec ~ dgamma(0.001, 0.001) # precision
```

```
for (i in 1:18) # for each experimental plot
{
  pred[i] <- a + b[Refuge[i]]*Density[i] # linear
                                            predictor
  Mortality[i] ~ dnorm(pred[i], prec) # Mort. is
                                           norm.
                                           dist'd

}
}
```

Initial values for the Markov chain were:

```
list(a= 0, b=c(0,0,0), prec=100)
```

The data are given by:

Density[]	Refuge[]	Mortality[]
1.125	1	0.82
0.75	1	1.25
1.5625	1	1.74
1.625	1	1.97
2.5625	1	2.61
2.875	1	3.41
0.1875	2	0.79
0.5625	2	1.58
1.4375	2	1.88
1.8125	2	2.29
2.8125	2	2.29
4.75	2	3.16
3.6875	3	1.33
1.25	3	1.12
1.1875	3	1.5
2.3125	3	1.5
1	3	1.56
1.75	3	1.64
END		

The value of the variable Refuge[] defines whether
the density of refuges is low, medium or high. Mortality[]
is the daily mortality rate expressed as a percentage.
There were six replicate plots for each of the three classes
of refuge density, with goby density varying within each class.

The simpler of the two possible models is given below, with the same slope coefficient for each of the three refuge densities.

```
model
{
  a ~ dnorm(0, 1.0E-6) # intercept
  b[1] ~ dnorm(0, 1.0E-6) # effect of density at
                          resource level 1 (low)
  b[2] <- b[1] # effect at level 2 (medium) same as
               level 1
  b[3] <- b[1] # effect at level 3 (high) same as
               level 1
  prec ~ dgamma(0.001, 0.001) # precision
  for (i in 1:18) # for each experimental plot
  {
    pred[i] <- a + b[Refuge[i]]*Density[i] # linear
                                           predictor
    Mortality[i] ~ dnorm(pred[i], prec) # Mort. is
                                        norm. dist'd
  }
}
```

Because b[2] and b[3] are now logical nodes, they can no longer be assigned initial values. Thus, the following were used as initial values for the Markov chain, with NA symbolizing that these values are not assigned stochastically.

```
list(a=0 b=c(0,NA,NA), prec=100)
```

The model in which the slopes did not differ between the different refuges had a very much inferior DIC value compared to the more complex model (33.4 versus 23.8). The difference in DIC values suggests that the effect of goby density on mortality does indeed depend on the availability of refuges.

The second experiment

Forrester and Steele (2004) conducted a second experiment, in which the manipulation was repeated, but the study was also conducted on plots that were larger and smaller than those used in the original experiment. The aim was to determine whether the observed effects of resource availability and goby density depended on the spatial scale of the experiment.

In the second analysis, Forrester and Steele (2004) were not able to identify a statistically significant effect of resource availability on goby mortality. However, the results were suggestive of an effect and appeared to be consistent with those of the first (Box 6.11). In conducting their analysis, Forrester and Steele (2004) were forced to ignore the results of the previous experiment because they used frequentist analysis, despite the fact that some aspects of the second experiment were identical to the first. By using Bayesian statistics, the results of the first experiment can be used to help strengthen the conclusions drawn from the second.

As for the mulgara example (Box 6.7), it is useful to compare the DIC value of a model with informative priors with an equivalent model with uninformative priors. This helps to indicate whether the prior is consistent with the new data. Additionally, for the second goby experiment, models in which the effect of goby density on mortality depends on both the spatial scale of the experiment and the resource availability are compared with models in which only resource availability matters. The WinBUGS code for these models is presented in Box 6.12.

Box 6.12
ANCOVA using an informative prior

Forrester and Steele (2004) repeated the experiment described in Box 6.11, but also included plots that were smaller and larger than those used in the original experiment. If the size of the plots is important, then the regression coefficients for the effect of density on mortality will depend not only on resource availability but also the plot size that was used. This can be expressed in the ANCOVA as a three-way interaction between the effect of goby density, refuge density and size of the plot. The presence of a two-way interaction between goby density and refuge density had already been supported by the previous experiment, so this was also included in the analysis. The two-way interaction between goby density and plot size was excluded from the model to reduce the number of estimated parameters and because the focus was whether the plot size influenced the interaction between goby density and refuge density.

The variable to specify the size of the plots was included in the model by using two binary variables to indicate whether the plot was small or large. If the plot was neither small nor large (in which case both binary variables were equal to zero), then the plot was the same size as used in the first experiment. This form of coding for categorical variables is known as using dummy variables (see Quinn and Keough, 2002 for details).

The three-way interaction terms are included by modifying the effect of goby density on mortality, depending on the refuge density and the plot size. This leads to six additional estimated parameters. The code is given below, using the results of the first experiment to set the informative priors for this analysis.

```
model
{
  a ~ dnorm(1.01, 24.826) # intercept
  b[1] ~ dnorm(0.628, 56.70) # effect of density at
                                resource level 1 (low)
  b[2] ~ dnorm(0.488, 110.83) # effect of density at
                                 resource level 2
                                 (medium)
  b[3] ~ dnorm(0.180, 70.144) # effect of density at
                                 resource level 3
                                 (high)
  intnS[1] ~ dnorm(0, 1.0E-6) # interaction terms for
                                 small plots at 3
                                 refuge densities
  intnS[2] ~ dnorm(0, 1.0E-6)
  intnS[3] ~ dnorm(0, 1.0E-6)
  intnL[1] ~ dnorm(0, 1.0E-6) # interaction terms for
                                 large plots at 3
                                 refuge densities
  intnL[2] ~ dnorm(0, 1.0E-6)
  intnL[3] ~ dnorm(0, 1.0E-6)
  prec ~ dgamma(0.001, 0.001) # precision
  for (i in 1:25) # for each of the 25 plots
  {
    pred[i] <- a + b[Refuge[i]]*Density[i] +
    intnS[Refuge[i]]*Small[i]*Density[i] +
    intnL[Refuge[i]]*Large[i]*Density[i]
```

```
    Mortality[i] ~ dnorm(pred[i], prec)
  }
}
```

Models with uninformative priors can be used by changing the prior distributions for the variables a and the b[]'s to, for example, dnorm(0.0, 1.0E-6). Models in which there is not an effect of plot size can be considered by setting all the interaction terms (intnS[] and intnL[]) to zero (e.g., intnS[1] <- 0, etc.).

The parameter estimates for the models with informative and uninformative priors are given in the tables below. Not surprisingly, by using informative priors for the parameters, their precision has improved (the standard deviations of the posterior distributions have decreased). However, the precision of the interaction terms is also improved by using informative priors for the other parameters. This means that by using the prior information, we can be more certain about the magnitude of the effect that plot size has on the experiment. The 95% credible intervals for the interaction terms encompass zero and the means of the posterior distributions are approximately equal to zero, indicating that there is little evidence that the plot size has influenced the results.

Parameter estimates for the ANCOVA model using uninformative priors for a and the b[]'s.

Node	Mean	sd	95% CI
a	0.3651	0.7086	−1.044–1.762
b[1]	1.886	0.6441	0.6099–3.168
b[2]	0.4665	0.3624	−0.2495–1.188
b[3]	0.4905	0.5636	−0.6276–1.6
intnL[1]	0.8001	1.736	−2.651–4.256
intnL[2]	0.4327	1.114	−1.784–2.644
intnL[3]	0.04674	0.5961	−1.136–1.229
intnS[1]	0.2165	0.8902	−1.539–1.983
intnS[2]	0.16	0.3899	−0.6134–0.9301
intnS[3]	−0.2155	0.5131	−1.228–0.7992

Parameter estimates for the ANCOVA model using informative priors for a and the b[]'s derived from the first experiment.

Node	Mean	sd	95% CI
a	1.001	0.1814	0.645–1.357
b[1]	0.6918	0.1302	0.4362–0.9473
b[2]	0.4604	0.0894	0.2855–0.6363
b[3]	0.1834	0.1156	−0.04332–0.4115
intnL[1]	0.9717	1.553	−2.086–4.058
intnL[2]	−0.09684	1.053	−2.192–1.988
intnL[3]	0.1056	0.3755	−0.6383–0.8441
intnS[1]	0.8003	0.7307	−0.6279–2.256
intnS[2]	−0.09628	0.302	−0.6957–0.494
intnS[3]	−0.06704	0.1759	−0.4135–0.2788

The DIC values (see table below) suggest that the two models without the three-way interaction terms (intnS and intnL) are the most parsimonious. Thus, there is good support for the two models in which the plot size does not affect the results. Of these two, the model with the uninformative prior (DIC = 76.7) has a slightly smaller DIC than the model with the informative prior (DIC = 77.8) but the difference is sufficiently small that the two are indistinguishable.

Model	Prior	DIC
with interaction term	uninformative	90.4
without interaction term	uninformative	76.7
with interaction term	informative	87.8
without interaction term	informative	77.8

Finally, the model in which there is no effect of plot size can be compared to one in which there is additionally no interaction between refuge density and goby density (i.e., in which all b[]'s are the same). The DIC for the latter model (85.4, using uninformative priors) is substantially greater than that for the model that includes the interaction (76.7), confirming the result of the first experiment where the effect of con-specifics on mortality depended on the density of refuges.

The analyses indicate that there is little support for the models that include the three-way interaction terms. Thus, the effects of competition and refuge density appear to be consistent for all plot sizes. Further, the influence of refuge density on the effect of intra-specific competition appears to be similar in both experiments.

Log-linear models for contingency tables

Contingency tables are commonly used to analyse relationships among variables, with the frequency of occurrence presented in cells for different combinations of factors. For example, French and Westoby (1996) presented a contingency table showing the number of plant species that have each possible combination of dispersal mechanism and regeneration strategy (Table 6.2). One of their questions was whether there is an association between these two factors because plants with seeds that are dispersed by vertebrates are predicted to be more likely to regenerate vegetatively compared to ant-dispersed species (French and Westoby, 1996).

These models can be analysed using chi-squared goodness-of-fit measures (Fowler *et al.*, 1998; Quinn and Keough, 2002). However, log-linear models represent these relationships with greater flexibility (Agresti, 1990; Quinn and Keough, 2002). In these models, the logarithm of the expected frequency is a linear function of the factors, with the factors treated as explanatory variables analogous to those of ANOVA (e.g. Box 6.5). Therefore, the expected number of species of plants (n_{ij}) would depend on the effects of the dispersal mechanism i (d_i) and regeneration strategy j (r_j), and the interaction between the two (b_{ij}):

$$\ln(n_{ij}) = a + d_i + r_j + b_{ij},$$

Table 6.2. *Number of ant- and vertebrate-dispersed plant species with seed and vegetative regeneration (French and Westoby, 1996).*

	Ant	Vertebrate	Total
Seed only	25	6	31
Vegetative	36	21	57
Total	61	27	88

where a is the intercept term. This equation can be re-expressed as:

$$n_{ij} = \exp(a) \times \exp(d_i + r_j + b_{ij}).$$

The expected proportion of species with dispersal mechanism i and regeneration strategy j will equal $n_{ij}/\Sigma n_{ij}$. The intercept term ($\exp(a)$) cancels out in this expression for the expected proportion, so it is excluded from the analysis.

An association between the variables is indicated when one or more of the interaction terms (b_{ij}) differ from zero. In contingency tables there are $(R-1)(C-1)$ estimable interaction terms, where R and C are the number of rows and columns in the contingency table. Therefore, the contingency table of French and Westoby (1996) has a maximum of one non-zero interaction term ($R = 2$, $C = 2$).

Given that the response variable (e.g. the number of species in Table 6.2) is a non-negative integer, a Poisson distribution might be appropriate for the analysis. However, the numbers of species in the cells of the contingency table are not independent, because they are constrained to equal the total number of species (88 in this example, Table 6.2). A multinomial model (see Box 3.14 and Appendix B) is able to accommodate such constraints (Box 6.13). The analysis suggests a possible positive association as predicted.

Agresti (1990), Sokal and Rolf (1995) and Quinn and Keough (2002) provide further details of analysing contingency tables. Albert (1997) provides a Bayesian perspective, including the analysis of models in which the number of non-zero interaction terms is specified a priori.

Box 6.13
Analysis of contingency tables

French and Westoby (1996) examined the relationship between the occurrence of vertebrate dispersal and vegetative reproduction in plants. Often the variable of interest when analysing contingency tables is how the relative probabilities compare. For example, we can analyse the plants with seeds that are dispersed by vertebrates to determine the relative proportion that have vegetative reproduction. This relative proportion can be expressed as odds, the proportion of the plants with vegetative regeneration divided by the proportion that regenerate only by seed. A positive association between vertebrate dispersal and vegetative reproduction is indicated if the odds for vertebrate-dispersed species is greater than the odds for ant-dispersed species (i.e. if the ratio of these odds is greater than one).

The WinBUGS code for the analysis is:

```
model
{
  for (i in 1:4) # for each of the 4 cells in the table
  {
    # calculate relative frequency of species
    log(p[i]) <- k1[Dispersal[i]] +
    k2[Regeneration[i]] + k12[Dispersal[i],
    Regeneration[i]]
  }
  pp <- sum(p[1:4]) # sum of the 4 values of p[]
  for (i in 1:4) # for each of the 4 cells in the table
  {
    pr[i] <- p[i]/pp # re-scale so pr[] are
                        probabilities
  }
  odds_vert <- pr[4]/pr[2] # odds for vertebrate-
                              dispersed species
  odds_ant <- pr[3]/pr[1] # odds for ant-dispersed
                              species
  odds_ratio <- odds_vert/odds_ant # odds ratio
  N <- sum(f[1:4]) # number of species — 88 in this
                      example
  f[1:4] ~ dmulti(pr[], N) # numbers drawn from a
                              multinomial distribution
  k1[1] <- 0 # parameters for reference classes set to
                zero
  k2[1] <- 0
  k12[1,1] <- 0 # k12[]'s are the interaction terms
  k12[1,2] <- 0
  k12[2,1] <- 0
  k1[2] ~ dnorm(0, 1.0E-6) # effect of dispersal
                              strategy
  k2[2] ~ dnorm(0, 1.0E-6) # effect of regeneration
                              mechanism
  k12[2,2] ~ dnorm(0, 1.0E-6) # interaction term
}
```

The above code models regeneration mechanism and dispersal mode as explanatory variables using reference classes (data and initial values are given on the book's web site). The posterior distribution of the odds ratio has a 95% credible interval of [0.93, 7.7] and a mean of 3.0. This suggests a possible positive association between vertebrate dispersal and vegetative reproduction, although the credible interval encompasses one near its lower bound. This association is also reflected in the interaction term (k12[2,2]), which has a 95% credible interval of [−0.08, 2.1] that includes zero near its lower bound.

Quinn and Keough (2002) obtained a 95% confidence interval [0.86, 6.9] for the odds ratio, which is similar to the 95% credible interval. Based on a non-significant test of the null hypothesis of independence ($P = 0.09$), Quinn and Keough (2002, p. 383) concluded that 'we have no evidence to reject the [null hypothesis] of independence'. This is despite the observed association being positive and consistent with that predicted. Null hypothesis testing can trap researchers into concluding that a non-significant result means there is no evidence for an effect.

Concluding remarks

This chapter extended the regression analyses of the previous chapter to include categorical explanatory variables. The result is statistical models that include ANOVA, ANCOVA and log-linear models of contingency tables. Some of the examples demonstrate the use and evaluation of prior information. Other examples that use uninformative priors demonstrate the numerical congruence between frequentist and Bayesian analyses.

Regression analysis, ANOVA and ANCOVA are closely related. They are based on models of how the expected value of a variable changes as a function of explanatory variables, and models the deviation of the data around this expectation by using a specified probability distribution. Bayesian analyses permit flexibility in which distributions are used and arbitrarily complex forms for functional relationships. Therefore, typical assumptions of linearity, equal variances and normal distributions can be relaxed and modified as required.

Case studies

The previous chapters described some of the main forms of statistical modelling that ecologists are likely to encounter. The following chapters build on these by describing more detailed examples and more complicated forms of data analysis. These case studies are examples of the complexity that can be included in Bayesian analyses with relative ease.

7

Mark-recapture analysis

Obtaining estimates of survival and fecundity rates is a fundamental aspect of population ecology, because these parameters, along with dispersal rates, control changes in local population sizes. The usual way of monitoring fecundity and survival rates is to identify individuals in some manner and then monitor them over time. If the individuals are easily monitored, then researchers will know when they die or breed, and it is easy to calculate the required parameters. For example, one could estimate survival rates based on the proportion of individuals that survive a given period.

It is usually difficult to monitor the survival and reproduction of individuals precisely. It may be possible to know the exact fate of many plants, but even some plants may not be apparent for several years before reappearing with above ground parts (e.g. terrestrial orchids). For animals, individuals routinely go missing, and it is difficult to be sure whether an unobserved individual is alive or dead. When estimating annual survival probabilities, we need to also estimate the probability of detecting an individual given that it is alive.

Methods

Mark-recapture methods have been developed to accommodate the chance that an undetected individual is not dead but has simply been overlooked (Lebreton et al., 1992). These methods are closely related to those used to address the issue of detectability (Chapter 5; Mackenzie et al., 2002). Mark-recapture methods work by specifically modelling the probability of observing each individual as a product of both its survival rate and the probability of observing the individual given that it has survived.

For example, consider a study that lasts for three years. An individual is observed in its first year (when first captured) and then observed in the second, but not in the third. By observing the individual in the second year, we know it was alive up to that point, but its subsequent survival is uncertain. The individual might not be seen because it is dead or because it is alive and undetected.

Alternatively, consider an individual that is seen in the first year, is not seen in the second, but is seen again in the third. In this case, we know that the individual was alive in the second year, but that it was not detected because of imperfect detection rates. If such re-appearances occur infrequently, then we have evidence to suggest that detection rates are relatively high, because individuals are usually being seen when they are known to be alive. However, if they occur frequently, then we can be reasonably sure that detection rates are low and that the number of individuals seen in a given year is substantially less than the number of individuals that are likely to be alive. In this case, estimates of survival, for example, would be underestimated unless the chance of detection was taken into account.

There are two ways that mark-recapture models can be analysed in WinBUGS. In the first, the process of survival and resighting are described explicitly. In the second, the likelihood of the string of observations for each individual is calculated for the given model, and the string of observations for each individual is treated as a Bernoulli random variable (using the so-called 'ones trick' in WinBUGS). Both produce the same parameter estimates but have different advantages. The former is more easily coded in WinBUGS and provides a more intuitive description of the model. The latter can utilize more compact descriptions of the data and permit calculation of DIC values within WinBUGS. Both will be illustrated here, beginning with the more intuitive version of the code.

First, we consider the observation history of each individual. If observations in each period are coded as ones and absences as zeroes, then the observation history can be represented as a series of Bernoulli events. Given that an individual is alive in one year, its survival to the next year can be treated as a Bernoulli event with probability equal to the annual survival rate.

Then, given that it is alive, the probability of it being observed is equal to its detectability (or resighting rate). However, if an individual is dead, then the resighting rate for that individual is zero. Therefore, we can simulate the survival from year to year, recording whether the individual is alive or not, and then simulate the resighting of individuals, with the resighting rate conditional on the animal being alive. The mark-recapture analysis of this intuitive form of the model is shown in Box 7.1.

One of the disadvantages of the intuitive form of the model is that WinBUGS does not calculate DIC values for the model. This can be overcome by specifying the likelihood of the capture history of each

---

Box 7.1
Mark-recapture analysis

The mark-recapture model below uses the variable `alive[i, j]` to record whether individual `i` is alive in year `j`. For the year in which the individual was first recorded, `alive[i, j]` is equal to 1 because we know that the individual was alive at that time. Subsequent survival is determined stochastically.

The probability of an individual being alive in the current year (`palive[i,j]`) is equal to the annual survival rate (surv) if it was alive last year, and zero if it is dead. Thus, `palive[i,j] <- surv * alive[i, j-1]`. Then, its status in the current year is determined by drawing it from a Bernoulli distribution. Similarly, for the resighting history, the individual can only be observed if it is alive. Thus, the probability of sighting an individual is equal to the annual resighting rate if it is alive and zero if it is dead (`psight[i,j] <- resight * alive[i,j]`). The actual observation of the individual is determined stochastically.

The remainder of the code simply specifies the priors for the survival and resighting rates.

```
model
{
  for (i in 1:N) # for each bird
  {
    alive[i, First[i]] <- 1 # 1 means it is alive the
                              first time it was seen
    for (j in First[i]+1:Years) # for each year after
                                   the first
    {
      #  palive = prob of remaining alive (=0 if
        previously dead, =surv if alive)
        palive[i,j] <- surv * alive[i, j-1]
      # determine whether it is alive
```

```
    alive[i,j] ~ dbern(palive[i,j])
    # probability of resighting depends on whether it
      is alive
    psight[i,j] <- resight * alive[i, j]
    # actual resighting determined randomly
    Y[i, j] ~ dbern(psight[i,j])
    }
  }
# Uninformative priors for survival and resighting
  rates
surv ~ dunif(0, 1)    # uninformative
resight ~ dunif(0,1) # resighting rate -
                       uninformative
}
```

The above code requires that the data for each individual includes its resighting history over all years (Y[]) and the year that it was first observed (First[], see website for code). Further, it is necessary to specify the initial values for the variable alive manually. To achieve this, it is simplest to assume that all the individuals are alive in all years after they were first sighted. The choice of these initial conditions does not influence the samples from the Markov chain.

individual, and using the 'ones trick' in WinBUGS to generate the required distribution (see Box 7.2).

The likelihood for the mark-recapture model needs to account for the possibility that individuals are either alive or dead. For the period from the first until the last sighting, it is known that the individual is alive, so all absences can be safely assumed to be failures of detection. Thus, for this period, the likelihood is equal to the probability of survival over the time period multiplied by the probability of obtaining the given detections and absences. The former part of this is simply the product of the survival rates over this time period. If it is assumed that survival is constant, then this product is equal to the survival rate (s) raised to the power of t_1, where t_1 is number of years from the first sighting until the last.

If, in this time, the individual is sighted d times (not including the first time it was sighted, so it was not sighted t_1-d times), then the likelihood for the detection component is equal to $r^d(1-r)^{t_1-d}$, with r being the

---

Box 7.2
The 'ones trick' in WinBUGS

WinBUGS contains a wide range of in-built probability distributions. However, it is common that other distributions might also be required. These can be incorporated in WinBUGS, if their likelihood function can be calculated, by using the so-called "ones trick" (see also the WinBUGS user manual).

The 'ones trick' works by first calculating the likelihood of each datum explicitly. For example, consider a Poisson random variable. The likelihood can be calculated from the probability distribution function for Poisson random variables. All that is required is that the likelihood is proportional to the probability. Thus, the likelihood of an observation Y is:

$$L(\gamma) = e^{-\lambda} \lambda^{\gamma} / \gamma!$$

For numerical convenience (WinBUGS calculates $\ln(Y!)$ rather than $Y!$) and better precision, the above likelihood can be written as:

$$L(\gamma) = \exp[-\lambda + \gamma \ln(\lambda) - \text{in}(\gamma!)]$$

For discrete distributions, these likelihoods can be considered as the probability of obtaining the observed data. For continuous probability functions, these likelihoods are probability densities and can be greater than one. When using the 'ones trick' in such cases, they need to be re-scaled (multiplied by a constant) to ensure that they are not greater than one (see the WinBUGS manual).

The tricky part of the "ones trick" is that each data point is thought of as the outcome of a Bernoulli trial with the probability success determined by the likelihood. A dummy variable is introduced that specifies the outcome of the Bernoulli trial. This variable takes a value of one for all observations; it specifies that the observed data did actually occur. The code below illustrates the 'ones trick' for a Poisson distribution, using the same model and data as in Chapter 3 (Box 3.4).

```
model
{
  for (i in 1:10) # for each of the 10 data points
```

```
{
  p[i] <- exp(-m + y[i]*log(m) ± logfact(y[i]))
  # calculate the likelihood of the data (y[i]),
     given the mean m
  Dummy[i] <- 1 # specify that the data were observed
  Dummy[i] ~ dbern(p[i]) # treat this observation
                              as a Bernoulli outcome
}
# prior for m
  m ~ dlnorm(0.0, 1.0E-6)  # a broad uninformative
                             prior for the mean
}
```

Because Dummy[i] is set equal to 1, WinBUGS tends to generate values of p[i] that are as large (i.e. as close to 1) as possible.

A total of 100 000 samples after discarding the first 10 000 produced an estimate of 2.5, with 95% credible intervals of 1.6–3.6. The original result had the same mean and credible interval (Box 3.4). Thus, the two approaches produce equivalent results, although the use of WinBUGS' in-built functions (e.g. dpois) is usually more efficient. However, the 'ones trick' and the related 'zeroes trick' (see the WinBUGS user manual) are useful when the desired distribution is not available.

annual resighting probability conditional on survival Thus, for the first period (from the first sighting until the last), the likelihood is equal to:

$$L_1 = s^{t_1} r^d (1 - r)^{t_1 - d}.$$

For the period following the last sighting, the individual may be alive or dead. If it died, it could have died in any one of the periods following the last sighting up to the final period of observation. If there was a single year in which the individual was not seen, then the likelihood for this period of time would be equal to the probability it died within that year plus the probability that it survived and was not seen. Thus, the likelihood would equal $(1-s) + s(1-r)$. If the individual was not seen for two years, then the likelihood would equal the probability it died in the first year $(1-s)$, plus the probability it survived the first year and was not seen but died in the second, plus the probability it survived both years and was not seen in either. Thus, the likelihood for the second component is equal to $(1-s)+s(1-r)(1-s)+s^2(1-r)^2$.

More generally, if the individual has not been seen for t_2 years, then the likelihood is equal to:

$$L_2 = (1 - s) \sum_{i=1}^{t_2} [s(1 - r)]^{i-1} + [s(1 - r)]^{t_2}.$$

The likelihood for the entire sighting history is equal to the product of L_1 and L_2. The WinBUGS code for this analysis is shown in Box 7.3.

Box 7.3
Explicit calculation of the likelihood for mark-recapture models

The code is given below for analysing a mark-recapture model with constant resighting and survival rates. This uses the "ones trick" by explicitly calculating the likelihood of the data. Although less intuitive than the code in Box 7.1, it produces equivalent parameter estimates and permits the calculation of the DIC value within WinBUGS. An additional advantage is that the data can be condensed by counting the number of individuals with each of the different resighting histories. In the case of female European dippers, this reduced the data file for the resightings from 130 lines for individuals to 28 lines for the different observed resighting histories. The main advantages of this reduction are the reduced scope for errors when entering the data and the faster evaluation of the model in WinBUGS. Additionally, it is no longer necessary to provide the relatively cumbersome initial values for the variable `alive`.

```
model
{
  for (i in 1:N)
  {
    # number of years in which there were resightings
    resightings[i] <- sum(Y[i, First[i]:Last[i]])-1
    # likelihood up to last resighting
    L1[i] <- pow(surv, Last[i]-First[i]) * pow
    (resight, resightings[i]) * pow(1-resight,
    Last[i]-First[i]-resightings[i])
```

```
# L2a's are for the likelihoods for contingency of
  death in each year since last resighting
L2a[i, Last[i]] <- 0 # this term is necessary in
                          case Last[i]=Years
for (j in Last[i]+1: Years)
{
  L2a[i,j] <- pow(surv*(1-resight), j-Last[i]-1)
}
# L2b's are the likelihoods for survival since
  last resighting
L2b[i] <- pow(surv*(1-resight), Years-Last[i])
# L's are the overall likelihood - prob of surv up
  to last sighting times the sum of all possible
  contingencies (death or survival over the next
  years)
L[i] <- L1[i] * ((1-surv)*sum(L2a[i, Last[i]:
Years]) + L2b[i])
# uses a dummy variable of ones as the ''data'' -
  ''the ones trick'' - see WinBUGS manual
Dummy[i] <- 1
phi[i] <- pow(L[i], n[i]) # likelihood for n[i]
                               individuals
                               with this
                               sighting history
Dummy[i] ~ dbern(phi[i])
  }
# Priors
  surv ~ dunif(0, 1) # annual survival - uninformative
  resight ~ dunif(0,  1)   #  resighting  rate   -
                             uninformative
}
```

The mark-recapture model is applied in Box 7.4 to data on European dippers, an aquatic passerine for which there is a data set that is used extensively for illustrating and teaching mark-recapture analysis. A model with an uninformative prior is compared to one with an informative prior based on the regression of survival of passerines versus body weight (Box 6.6).

Box 7.4
Mark-recapture analysis of female
European dippers

Here we analyse the annual survival of European dippers (*Cinclus cinclus*), using data from Marzolin (1988) obtained over seven years for 130 individuals (McCarthy and Masters, 2005). With an uninformative prior, the posterior distribution for survival rate has a mean of 0.55 with 95% credible interval of 0.48−0.62.

However, in the previous chapter, by using information on body mass and annual survival for other European passerines, we were able to predict the annual survival of European dippers using only body mass (Box 6.6). This prediction can be used as an informative prior, by using a normal distribution with mean of 0.57 and standard deviation of 0.073 (precision = 187). With this prior, the posterior distribution for annual survival is 0.56 with 95% credible interval of 0.49−0.62. The information on body weight barely improved the estimate because the data are much more informative than the prior. The DIC values for these two models are very similar (344.3 and 343.9 for the uninformative and informative priors) because the data dominates the posterior distribution and both priors are consistent with the data.

The potential value of the prior can be examined by repeating analysis but only using the first 3 to 6 years of data. This illustrates how the prior can contribute to the estimate when fewer data are available. If only 3 years of data were available, the 95% credible interval when prior information is ignored had a width of 0.59. This is more than double that obtained when body mass is used (0.26), illustrating the increased precision provided by the prior. If prior information is ignored, 5 years of data collection would be required to obtain a more precise estimate than the one based on body mass alone (Fig. 7.1). Therefore, at the start of the mark-recapture study, the information on body mass is equivalent to 4 or 5 years of field data. In contrast, the width of the credible interval is 0.138 after 6 years when using prior information and is 0.141 after 7 years without prior information. By the end of the 7 year study, the improvement in precision by using the prior information is equivalent to adding one year of data in the mark-recapture analysis. Thus, using prior information derived from body mass is a very inexpensive way of adding precision to the study. In this example, the

prior information was worth between one and five years of field data
at the cost of reviewing and analysing data in the literature.

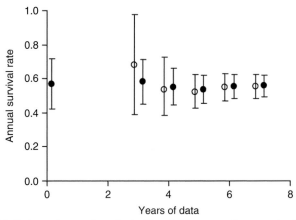

Fig. 7.1 Estimated annual survival of female European dippers using 7 years
of data and an uninformative (open symbols) and informative (closed
symbols) prior. The informative prior was based on a regression of annual
survival of European passerines versus body weight. The circle is the mean
of the posterior distribution and the bars are 95% credible intervals
(modified from McCarthy and Masters, 2005).

It is relatively easy to modify these models to permit annual variation
in the resighting or survival rate. In this case, rather than raising survival,
for example, to the power of t to provide the probability of survival over
t years, it would be necessary to calculate the product of all values of s_i
over the relevant years, where s_i is the survival rate in year i.

8

Effects of marking frogs

The analysis in Chapter 7 used data from individually identified animals to determine annual survival rates. Individual identification is usually achieved by marking the animals in some way. Birds are often marked by placing colour bands around their legs, with different colour combinations used for each individual. In this way, individuals may be recognized without the need for recapture.

An assumption of the analyses is that the marking method does not influence the animals. If the marking method does influence the animal, it can have consequences for both the conclusions drawn from the study and perhaps ethical implications. Some marking methods can have remarkable influences. For example, red colour bands in birds can influence reproductive performance (Burley et al., 1982; Hunt et al., 1997). Other impacts may be more severe. For example, penguins with flipper tags tend to have lower breeding probabilities and reproductive rates (Gauthier-Clerc et al., 2004).

The most common method of marking amphibians is to remove a unique combination of toes (or part thereof) to identify each individual. A total of eight toes from a full complement of 18 have been removed from some individuals in some studies, but almost 1000 individuals can be marked by removing up to only three toes (Hero, 1989; Waichman, 1992).

Not surprisingly, there has been some concern that the removal of toes could have adverse effects on the individuals (May, 2004). Various impacts have been reported from inflammation and infection of the wounded digits, to apparent reductions in the chance of recapturing the marked individuals. However, results are somewhat variable, with some authors reporting adverse effects while others did not find such effects.

One of the most uncertain and important aspects to the question of possible adverse impacts is that toe clipping may influence the return rate. The return rate is the product of the survival rate and the recapture rate (conditional on survival). If the return rate declines with increase in the number of toes clipped, then parameter estimates from mark-recapture studies are likely to be biased. Furthermore, there would be the distinct possibility that toe clipping increases the mortality rate of individuals.

A reduction in the return rates of toe-clipped frogs and toads have been reported in some cases (e.g. Clarke, 1972; Humphries, 1979; Williamson and Bull, 1996), but not in others (e.g. Lemckert, 1996; Williamson and Bull, 1996). Parris and McCarthy (2001) helped to resolve this apparent inconsistency by demonstrating that absences of statistically significant effects in some previously published studies could be attributed to a lack of statistical power rather than absences of actual effects. Based on fitted regression lines, return rates were estimated to decline by 6−18% for each toe removed after the first (Parris and McCarthy, 2001). However, this analysis did not provide meaningful confidence intervals for the estimate, or analyse how the impact of toe clipping might change with the number of toes removed.

Here, a Bayesian re-analysis of the data used by Parris and McCarthy (2001) is presented, illustrating a case study conducted by McCarthy and Parris (2004). The Bayesian analysis provides a relatively easy way to include more biologically relevant statistical models that actually estimate the effect of toe clipping on a per toe basis. The analysis permits the influence of toe clipping to increase or decrease with the number of toes removed, thereby demonstrating a relatively consistent effect of toe clipping on frogs and toads.

The data were obtained from four previously published studies of the influence of toe clipping on the return rate of frogs (see Parris and McCarthy, 2001). Williamson and Bull (1996) studied 1333 individuals of *Crinia signifera* with up to seven toes removed from each individual. *Crinia signifera* is a small, ground-dwelling frog from eastern Australia that grows up to 30 mm snout-vent length (SVL). Lemckert (1996) also studied *C. signifera*, with 306 individuals and between two and four toes removed from each individual. Clarke (1972) reported the effect of removing up to eight toes from 733 individuals of *Bufo fowleri*, and Lüddecke and Amézquita (1999) reported effects of toe-disk clipping on the return rate of 1307 individuals of *Hyla labialis*, with up to seven toe-disks removed from each individual. *Bufo fowleri* is a relatively large, ground-dwelling frog from the eastern United States (up to 80 mm SVL), while *H. labialis* is a medium-sized tree frog from the Colombian Andes (up to 55 mm SVL). Extra information was available in some of the studies, such as the return rate for different sexes and years, and different size and age classes. This extra information was used in the statistical analysis to account for some of the variation in the data. The statistical models investigated are described below.

Logistic regression

The original statistical model used by Parris and McCarthy (2001) was based on logistic regression. This had the advantage over previous analyses (such as correlation analysis and linear regression) of accounting for the binomial nature of the data; each individual either returned or it did not, and the differences in sample sizes for different numbers of toes clipped could be accommodated. Logistic regression (Chapter 5) relates the return rate of frogs to the number of toes clipped using the formula:

$$\ln(R(n)/[1 - R(n)]) = A + Bn,$$

where $R(n)$ is the expected return rate of frogs that have had n toes removed, B is the regression coefficient for the effect of toe clipping and A is a value that may include terms for other covariates (such as yearly differences in return rate). When $B < 0$, the return rate decreases with the number of toes removed.

The logistic equation illustrates one of the disadvantages of using logistic regression, because the variable of most interest (the change in the return rate for each toe removed) is not included explicitly in the equation. The change in return rate can be estimated by examining how the predicted return rate changes with each additional toe that is removed beyond the first. However, in the analysis of Parris and McCarthy (2001) this procedure would lead to 87 different estimates of the effect of toe clipping, which would be difficult to interpret. Parris and McCarthy (2001) determined the change in return rate with each toe removed by inspecting the fitted regression lines. This provided useful information on the effect of toe clipping, but there was some subjectivity in estimating the magnitude of the effect, making it impossible to place meaningful confidence intervals on the estimate.

Model A

An alternative model can be developed by assuming that the return rate changes by a constant proportion for each toe removed. If the removal of a toe causes a constant change in the return rate (m), then the return rate will equal $R(0) \times (1 + m)$ following the removal of one toe. If a second toe is removed, this return rate ($R(0) \times (1 + m)$) will be further changed and the return rate will equal $R(0) \times (1 + m)^2$. It follows that the return rate following the removal of some number of toes (n) will equal $R(0) \times (1 + m)^n$. If there is a reduction in return rate due to mortality, m will be negative, $-m$ will equal the proportion of frogs that die following the removal of each toe, and $1 + m$ will be the chance of surviving the removal of a toe. If toe clipping causes a behavioural response such as aversion to recapture, or migration away from the site of initial capture, then m would be interpreted as the change in the probability of recapture (for each toe) given that the frog is alive. Positive values of m would indicate that removing toes increases the return rate, perhaps by making the individuals less mobile and more likely to remain in the study area. A value of zero for m represents no effect of toe clipping. Thus, the first model used in the re-analysis (Model A) was:

$$R(n) = R(0) \times (1 + m)^n.$$

To account for different return rates for different types of frogs, different values of $R(0)$ were estimated for frogs of different sizes, sexes, or ages where such data were provided by the original authors. The analysis also distinguished between individuals caught in different years to account for annual variation in return rates. For the data of Williamson and Bull (1996), differences between juveniles, adult males and adult females in each of three years were considered. For the data of Clarke (1972), the analysis distinguished between large and small individuals. Differences among years and sex were examined for the data of Lüddecke and Amézquita (1999). Lemckert (1996) did not distinguish among different classes of *C. signifera*.

In addition to defining the return rate, it was assumed that the fate of each individual in the same class (i.e. individuals with the same number of toes clipped, of the same age, etc.) was determined independently of the other individuals. As a result, the number of frogs recaptured was drawn from a binomial distribution (Appendix B), with variance equal to $NR(1-R)$, where N is the number of frogs in the class that were marked and released, and R is their return rate.

Models B and C

Model A assumed that the influence of toe clipping was the same for each toe removed. However, the effect of removing a toe may be greatest for the first toe removed (diminishing impact), or the effect per toe may increase with each toe removed (increasing impact). It is possible to incorporate such a modification to the preceding equation. In this case, the effect of toe clipping may be expressed as:

$$R(n) = R(0) \times (1 + m_1) \times (1 + m_2) \times \ldots \times (1 + m_n),$$

where m_n is the change in return rate when removing the nth toe. A linear function modelled the change in m, so $m_n = a + bn$. More complex functions could be used, but this choice has the advantages of simplicity and ease of interpretation. The estimated parameters for the chosen function were used to determine if there was an increasing or diminishing impact. In addition to analysing the data sets separately (Model B), the data were also pooled from the four studies and the values of m_n estimated under the assumption that the parameters a and b were the same for all studies (Model C).

The WinBUGS code for conducting the analysis of Model B is given in Box 8.1. A total of 100 000 samples from the posterior distributions for each of the models was generated after discarding the initial 10 000 samples as a 'burn in'. The mean of each of the parameters was calculated, as was the 2.5th and 97.5th percentiles of the distribution to represent a 95% credible interval.

For the analysis assuming a constant effect of toe clipping for each toe removed (Model A), the results were broadly consistent with those of Parris and McCarthy (2001), with return rates reduced by approximately 4–11% for each toe removed (shown previously in Fig. 2.3). There was strong evidence for a negative effect of toe clipping in the studies of Williamson and Bull (1996), Lüddecke and Amézquita (1999) and Clarke (1972), because the upper limit of the 95% credible intervals were less than zero. This was equivalent to the conclusion of Parris and McCarthy (2001), who determined that the observed decline (or a larger decline) was unlikely to have occurred if the number of toes removed did not affect the return rate. For the study of Lemckert (1996), the negative mean provided some evidence that toe clipping reduces return rates (Fig. 2.3), although it is possible that there is no effect or that toe clipping increases return rates (the credible interval encompassed zero). The wide credible interval for

Box 8.1
Analysing return rates with a non-linear model

The models of the effect of toe clipping on frogs were analysed using uninformative priors. It would be possible to incorporate reports of infection, anecdotal accounts, and intuition that suggest that toe clipping does not increase return rates. In most cases, the results were not sensitive to this choice because the data did not indicate a possible positive effect of toe clipping on return rates. The upper bound on the change in return rate (m) was set to ensure that the return rate for the maximum number of toes clipped was not greater than 1; the lower bound on m was -1. The prior distribution was uniform between -1 and 1 for a, and between -0.2 and 0.2 for b. These limits for the prior distributions did not constrain the posterior distributions.

In addition to the models described in the text (Models A, B, and C), the logistic regression models used by Parris and McCarthy (2001) were analysed, using uninformative normal priors (mean of 0 and standard deviation of 1000) for the regression coefficients. The fit of the various models was compared using DIC (Spiegelhalter *et al.*, 2002; Chapter 4).

The following is the WinBUGS code used to analyse Model B. The variable change is modelled as a linear function of the number of toes clipped. The analysis is 'centred' around three toes clipped (change = a + b*(Toes-3)) to reduce autocorrelation of the parameter estimates and correlation among the parameters, particularly a and b (Box 5.8). A model using change = a + b*Toes produces identical results, but the sampling is less efficient.

```
model
{
  for(i in 1: NGROUPS) # a group is a collection of
                         individuals with the same
                         characteristics (same
                         study, sex, number of toes
                         clipped, etc.)
  {
    y[i] ~ dbin(RR[i], n[i])   # no. returning is a
                                 binomial sample
    RR[i] <- RR0[classID[i]]*relRR[study[i],
    Toes[i]]
```

```
  # return rate depends on the class and no. of toes
    removed
  # RR0[i] is return rate of class i if no toes
    removed
  # relRR[i, j] is return rate of study i with j toes
    removed, relative to return rate if no toes
    removed RR0
}
for (j in 1:NSTUDIES) # the relationship between
                        toes clipped and change in
                        return rate is allowed to
                        vary among studies in
                        Model B
{
  change[j, 1] <- a[j] + b[j]*(1-3) # change in RR
                                     with  1  toe
                                     removed
  relRR[j, 1] <- 1 + change[j, 1]
  # relative return rate if one toe removed
  for (i in 2:MAXTOES)
  {
    change[j, i] <- a[j] + b[j]*(i-3) # change in RR
                                       with i toes
                                       removed
    relRR[j, i] <- relRR[j, i-1]*(1+change[j, i])
    # relative return rate if i toes removed
  }
}
# PRIORS - uninformative
  for (i in 1: NSTUDIES)
  {
    a[i] ~ dunif(-1, 1)
    b[i] ~ dunif(-0.2, 0.2)
  }
  for(i in 1: NCLASSES)
  {
    RR0[i] ~ dunif(0, 1)
  }
}
```

this study was consistent with its low statistical power (Parris and McCarthy, 2001).

As discussed in Chapter 2, by using credible intervals to present the results, the predicted impacts of toe clipping can be compared to values that might be considered biologically important, not just statistically significant (Fig. 2.3). For example, we can be confident that the reduction in return rate was greater than 0.03 (equivalent to one frog not returning due to toe clipping for every 33 toes removed) for three of the four studies. In the other study (Lemckert, 1996), the results were also consistent with an impact of at least 1 in 33.

The analysis in which the effects of toe clipping were permitted to change with each toe removed (Model B) provided evidence for increasingly negative effects with each toe removed in the studies of Williamson and Bull (1996) and Lüddecke and Amézquita (1999) (Fig. 8.1). The study of Clarke (1972) suggested an increasing impact, although the possibilities of no impact or a declining impact could not be eliminated given the width of the credible intervals. The data from Lemckert (1996)

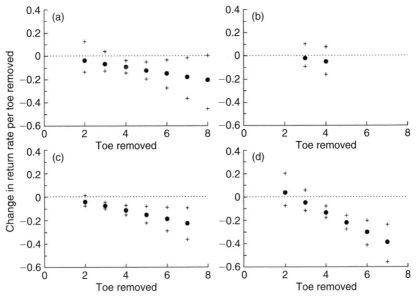

Fig. 8.1 Change in return rate for each toe removed assuming Model B for each of the four studies. The circles are the mean of the posterior distribution and the crosses are the limits of the 95% credible intervals (from McCarthy and Parris, 2004).

provided little insight into this particular question, with few data to indicate a trend and relatively wide credible intervals.

The influence of toe clipping on return rates appeared to be relatively consistent across the different studies, because the credible intervals for the different studies largely overlapped (Fig. 8.1). Thus, Model C seemed to be appropriate for the data. Removing the second toe was estimated to reduce return rates by approximately 3.5% ($m_2 = -0.035$), with the impact increasing to approximately 30% for the eighth toe ($m_8 = -0.3$) (Model C). These impacts were in addition to the effect of removing previous toes. Because the 95% credible interval for toe number two had an upper bound of approximately zero, it is possible that removing a second toe could have a negligible effect on the return rate. There are no animals in the data sets without toes removed, so it is not possible to estimate the impact of removing the first toe. Extrapolation suggests this impact may be small (Fig. 8.2), although this should be regarded as speculative because we cannot be sure that the relationship is even approximately linear beyond the range of the data.

Models B and C, in which the effect of toe clipping increased with each toe removed, fit the data better than Model A, based on the calculated DIC values (Table 8.1). These two models were largely indistinguishable on the basis of DIC, with a difference of one unit. Model A in turn provided a better fit than the original logistic regression model of Parris and McCarthy (2001).

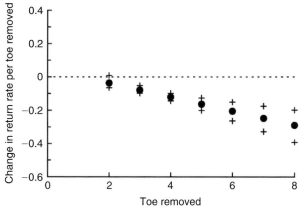

Fig. 8.2 Change in return rate for each toe removed assuming Model C. The circles are the mean of the posterior distribution and the crosses are the limits of the 95% credible intervals (from McCarthy and Parris, 2004).

Table 8.1. *Values for the deviance information criterion (DIC) indicating the goodness of fit of the original logistic regression of Parris and McCarthy (2001) and the three new models (Models A, B and C).*

Model	DIC
Logistic regression	532.6
A − constant effect of toe clipping	522.9
B − changing effect varying among studies	506.2
C − changing effect consistent across studies	507.2

The results of the re-analysis of published data on the effects of toe clipping on the return rate of frogs are consistent with those of Parris and McCarthy (2001). However, the re-analysis also suggests that apparent differences among previous studies in the effect of toe clipping on return rates may be due to the different number of toes that were removed from individual animals. The model with a consistent, linear change in the effect of toe clipping with each toe removed (Model C) demonstrated that the impact of clipping each toe increases as more toes are removed, corresponding to a rapidly compounding effect on the behaviour and/or survival of the marked frogs. Frogs have 18 toes in total, four on each of the fore feet and five on the hind feet. As well as assisting with balance and locomotion, the enlarged, adhesive disks on the toes of many tree frogs enable them to climb steep or vertical surfaces. It is perhaps not surprising that the fewer the toes a frog still possesses, the greater the effect of removing one more could have on its probability of return.

9

Population dynamics

WinBUGS is essentially a programming language that is particularly useful for generating a range of probability distributions. As such, it can be used to simulate a range of stochastic models. In ecology, some of the most common forms of stochastic models are those for predicting the viability of species (Shaffer, 1981; Gilpin and Soulé, 1986; Burgman *et al.*, 1993).

There are several advantages of using WinBUGS to simulate population dynamics. One is the broad range of probability distributions that is available. Another is that it is relatively easy to extend the simulation to consider how uncertainty in the parameter estimates for the model influences uncertainty in its predictions. This use of WinBUGS is illustrated here with a model of the dynamics of mountain pygmy possums.

Mountain pygmy possums

Mountain pygmy possums are small (adult body weight of 40−45 g) terrestrial possums inhabiting alpine regions of southeastern Australia. The species occurs in the Snowy Mountains in New South Wales, the Hotham/Bogong region in Victoria, and in the vicinity of Mt Buller in Victoria. Populations in these regions are variously fragmented. Typical breeding areas are boulder fields above the snowline (approximately 1500 m above sea level). The species is omnivorous, eating a range of invertebrates (Bogong moths in particular) and plant matter (Mansergh *et al.*, 1990; Smith and Broome, 1992). Breeding occurs in spring, with a single litter of four offspring being produced. Males migrate from the breeding areas by the end of summer, and typically over-winter in areas up to several kilometres away. Females are comparatively sedentary (Mansergh and Broome, 1994). Males and females hibernate over winter in crevices below the snow (Broome and Geiser, 1995; Körtner and Geiser, 1998).

The model developed here only considered females, because they appear to be the sex that limits population growth. Further, the actual age of individuals was ignored. This is partly because animals born in the previous summer are able to breed in the next breeding season and partly because changes in survival and fecundity rates with age are not known.

One of the most important aspects of a population model is to determine how population growth rates change with population size. For the mountain pygmy possum, females tend to be territorial, and there is likely to be competition for sites suitable for over-wintering. Therefore, density dependence was modelled using a Ricker function, in which the population growth rates decline as an exponential function with increasing population size.

The second aspect of the dynamics of the species to consider is that mountain pygmy possums typically have small population sizes. Consequently, there can be considerable variation in the population growth rate that only arises because of the chance birth and death of individuals. When population size is small, the population growth rate can vary even when the underlying chances of raising offspring and surviving do not change. This phenomenon is known as demographic stochasticity. The Poisson distribution is useful for incorporating demographic stochasticity (Akçakaya, 1990) and is used in this model.

Finally, environmental variation is included in the model by allowing the growth rate to vary annually. Such variation may occur, for example, because the level of snow cover fluctuates from year to year, as does the timing of its melting. This variation is likely to affect the ability of the mountain pygmy possum to hibernate over winter and conserve its fat stores. Additionally, the abundance of its food is also likely to vary annually. Environmental variation was included in the model by adding a normal random deviate to the expression for population growth rate. Thus, the expected population growth rate was given by:

$$\lambda_t = \exp(\alpha[1 - N_t/K] + \varepsilon_t),$$

where α is the maximum exponential growth rate, K is the equilibrium population size (the population size such that $\lambda_t = 1$ when $\varepsilon_t = 0$), N_t is the population size at time t, and ε_t is a normal random variable with mean zero and standard deviation σ.

The above model has three parameters that need to be estimated (α, K and σ). The data used to estimate these parameters and the associated WinBUGS code is given in Box 9.1.

Box 9.1
Estimating parameters of a population model

The data used to estimate the parameters of the model were obtained by Linda Broome and her various colleagues and assistants in the Snowy Mountains of New South Wales. Details of the data collection and the actual values are given in McCarthy and Broome (2000). Data were available from four sites for 12 or 11 years at each (giving data on 11 or 10 growth rates for each site). The data in this example are compact, so they are also provided here.

The priors for the model are relatively uninformative. The equilibrium population sizes were chosen to be uniform between 1 and 50, with a different equilibrium for each population. The maximum exponential population growth rate (alpha in the code below) was chosen to be uniform between zero and ln(3). The upper limit was chosen because females produce up to four young per year. Assuming the sex ratio is unity means that up to two female offspring are born. If all the mothers and female offspring survive (which is very unlikely), then the population size could at most triple each year. Therefore, alpha cannot conceivably be greater than ln(3). The code in other respects is straightforward. The model loops through the four populations and the years of data for each population.

```
model
{
  for (j in 1:4) # for each population
  {
    for(i in 1:T[j]) # for each year
    {
      # env stoch in population growth rate, drawn from
        normal
        ev[j, i] ~ dnorm(0, tau)
```

```
    # per capita growth rate is dens dep - Ricker
    Er[j, i] <- exp(alpha*(1 - N[j, i]/KB[j]) +
    ev[j, i])
    # lambda equal to number this year times per capita
    rate
    lambda[j, i] <- N[j, i] * Er[j, i]
    # number next year drawn from Poisson with mean
    lambda − demo. stoch.
    N[j, i+1] ~ dpois(lambda[j, i])
    }
  }
#PRIORS
# alpha is maximum exponential growth rate
  alpha ~ dunif(0, 1.0986)
# K's are carrying capacities of the 4 sites
  for (i in 1:4)
  {
    K[i] ~ dunif(1, 50)
  }
# st. dev. in growth rate due to env. stoch.
  sd ~ dunif(0, 0.5)
  tau <- 1/(sd * sd)
}
list(N = structure(.Data = c(
32, 28, 29, 39, 20, 24, 22, 35, 20, 36, 34, 40,
25, 25, 24, 24, 28, 15, 30, 36, 27, 31, 20, NA,
7, 11, 8, 14, 11, 5, 12, 6, 10, 12, 19, NA,
12, 16, 8, 8, 16, 11, 10, 6, 4, 10, 12, NA),
Dim = c(4, 11)),
T = c(11, 10, 10, 10))
```

The posterior distributions of the maximum growth rate
(alpha), the standard deviation and a representative equilibrium
population size (population 1) are shown in Fig. 9.1. The effect
of limiting alpha to be no more than ln(3) is seen by the sharp
cut-off at this value. The other parameters were not constrained
by their priors.

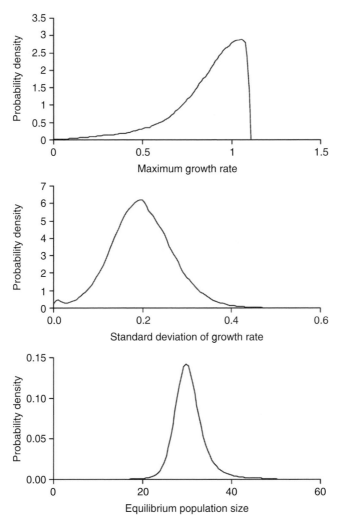

Fig. 9.1 Posterior distributions for the mountain pygmy possum model. The panels are for the maximum population growth rate, standard deviation of the growth rate and the equilibrium population size of population 1.

The code in Box 9.1 provides parameter estimates for the population model. The parameters could be exported from WinBUGS and used as input for a population model written in a separate package to predict future population sizes. However, it is also possible to do these predictions in WinBUGS. For illustration, predictions were made over a 20-year period and for a population with an equilibrium population size of 80 individuals. Of particular interest are the risk of declining to half the original population size (40 individuals or smaller) and the expected minimum population size (McCarthy, 1996; McCarthy and Thompson, 2001) within the 20-year period. The WinBUGS code for the analysis is given in Box 9.2.

Box 9.2
Posterior distributions for predictions of a population model

The following code predicts the risk of falling to 40 individuals or fewer (qe40) at some time within the next 20 years and the expected minimum population size (EMP). The first half of the code is identical to that in Box 9.1, and is used to estimate the parameters for the predictions. The second half of the code runs a specified number of iterations (e.g. ITS=1000) of the stochastic model to predict the risks of decline. The smallest population size within a 20-year period ($i = 1$ corresponds to time zero) is recorded using the ranked() function. A quasi-extinction (Ginzburg *et al.*, 1982) event is recorded if this minimum is less than or equal to 40. The minimum values averaged over the ITS iterations provide the expected minimum population size.

```
model
{
  for (j in 1:4) # for each population
  {
    for(i in 1:T[j]) # for each year
    {
    # env stoch in population growth rate, drawn from
      normal
      ev[j, i] ~ dnorm(0, tau)
    # per capita growth rate is dens dep - Ricker
      Er[j, i] <- exp(alpha*(1 - N[j, i]/K[j]) +
      ev[j, i])
    # lambda equal to number this year times per capita
      rate
      lambda[j, i] <- N[j, i] * Er[j, i]
    # number next year drawn from Poisson with mean
      lambda - demo stoch
      N[j, i+1] ~ dpois(lambda[j, i])
    }
```

```
}
#PRIORS
# alpha is maximum exponential growth rate
    alpha ~ dunif(0, 1.0986)
# K's are carrying capacities of the 4 sites
for (i in 1:4)
{
  K[i] ~ dunif(1, 50)
}
# st. dev. in growth rate due to env. stoch.
sd ~ dunif(0, 0.5)
tau <- 1/sd/sd
# the code above is identical to that in Box 9.1
# the following code is used to predict risks of
  decline
Kpred <- 80 # equilibrium population size
for (its in 1:ITS) # for each of the ITS iterations
{
  fem[its, 1] <- Kpred # initial pop size
  for (i in 2:21) # for a 20 year period
  {
    # environmental stochasticity
    envstoch[its, i] ~ dnorm(0.0, tau)
    # predicted population size
    pred[its, i] <- fem[its, i-1] * exp(alpha*
    (1 - fem[its, i-1]/Kpred) + envstoch[its, i])
    # demographic stochasticity - actual number
      drawn from Poisson
    fem[its, i] ~ dpois(pred[its, i])
  }
  # minimum pop size over 20 years
mini[its] <- ranked(fem[its, 1:21], 1)
                            # smallest pop recorded
q40[its] <- step(40-mini[its]) # =1 if the minimum
                              <= 40, 0 otherwise
}
EMP <- mean(mini[])/Kpred # EMP as a fraction of the
                            initial
qe40 <- mean(q40[]) # quasi-extinction risk
}
```

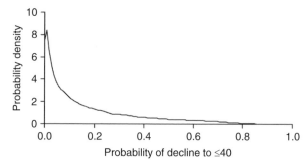

Fig. 9.2 Posterior distribution for the risk of a population of mountain pygmy possums falling to 40 individuals or fewer at some time within the next 20 years.

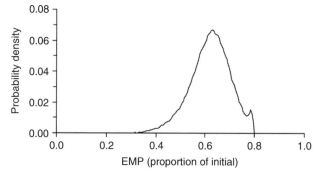

Fig. 9.3 Posterior distribution for the expected minimum population size of mountain pygmy possums within the next 20 years.

The posterior distribution for the risk of decline to 40 individuals (or fewer) was very wide, indicating considerable uncertainty in this estimate (Fig. 9.2). This is typical of efforts to assess risks (e.g. Ludwig, 1996). In this case, the risk of decline has a mode of almost zero, but the actual risk could be more than 0.75.

The expected minimum population size (EMP) can be predicted much more reliably than risks of decline. In this case, the posterior distribution for EMP (as a proportion of the initial population size) had a mean of 0.6, with the actual value almost certainly between 0.4 and 0.8 (Fig. 9.3).

10

Subjective priors

Because priors can have an important influence on the posterior distributions, their construction needs to be logical and repeatable. Subjectively generated priors, when combined with new data using Bayes' rule, indicate how a person's belief in parameter values should be updated to accommodate the new data. It is not surprising that such subjective treatments of knowledge raise concerns among scientists (e.g. Dennis, 1996). Is one person's subjective judgement a particularly valid basis for making scientific inferences?

Subjective judgement is useful for science in several circumstances. These include using subjective judgement to help interpret data, understanding how data can turn differences of opinion into agreement, and using subjective judgements coherently and explicitly in cases where time, resources and data are limited. Bayesian methods in these cases provide a more transparent treatment of that subjective judgement than either pretending it does not exist or considering the judgements qualitatively. An advantage of a Bayesian approach is that the subjective judgement can be combined logically with data. However, the use of subjective judgement is not inherently Bayesian; other approaches are available (Ayyub, 2001).

The process of eliciting subjective judgements should be documented and repeatable. Individual elicitation case studies differ in how questions are asked, how differences of opinion are handled, and how elicited information is used and combined with other sources of data (Ayuub, 2001; Burgman, 2005).

Eliciting probabilities

Elicited probabilities represent a person's beliefs in an event occurring. The simplest method of eliciting probabilities is to simply ask someone: 'What is the probability of an event occurring?' (Morgan and Henrion, 1990). Such questions can be difficult for people to answer, so various procedures have been developed to help that process. One procedure for eliciting a probability is to use an analogy to betting (Jaynes, 2003; Burgman, 2005). For example, if we wished to estimate the chance of an algal bloom, we could give a person a ticket that is worth a $100 reward if the algal bloom occurs. However, we also offer the person the chance of swapping the ticket for a reward that is payable regardless of the outcome but before knowing whether the bloom occurred or not. We begin by offering a small reward and increase it until the person is willing to sell their ticket. This point of indifference is the selling price (X_S) and the person's estimate of the probability of the algal bloom would equal:

$$p_S = X_S/100.$$

Of course, the scenario could be reversed by telling the person that we have a ticket that is worth $100 if the algal bloom occurs, and then ask if she would be willing to buy the ticket for a small value. This value is increased gradually until they are no longer willing to buy it. This buying price, X_B, can also be used to estimate the probability of the algal bloom

$$p_B = X_B/100.$$

Interestingly, p_B is usually less than p_S, which is just one example of how the format of the question can influence the answers (Burgman, 2005).

Handling differences of opinion

In eliciting probabilities it is common to ask the opinion of more than one person, and only rarely will they agree completely. At a very basic level, the result of elicitation depends on whom one asks, so the process of choosing subjects is important. People honestly perceive probabilities differently for several reasons, including their different experiences and level of expertise, their stake in the outcome, the magnitude of the event, and different tendencies towards overconfidence (Burgman, 2005). For example, someone concerned about adverse effects of an algal bloom might believe that the probability of a bloom is higher than someone whose management actions contribute to its occurrence.

Behavioural and numerical methods can help to resolve differences of opinion. Behavioural methods involve sharing of answers followed by clarification, re-evaluation and another round of elicitation. Numerical aggregation weights the different responses in some way to arrive at an average or bounded result. The weights assigned to different experts can be based on their ability to correctly assess probabilities that have been measured (Burgman, 2005). Regardless of the method used to aggregate results, differences of opinion may be real and should not be ignored. In some cases the diversity of views may be more instructive than the average view (Crome *et al.*, 1996).

Using subjective judgements

There are many different quantitative methods that use subjective judgement (Ayuub, 2001; Burgman, 2005). These include fuzzy arithmetic (Kaufmann and Gupta, 1985); information-gap theory (Ben-Haim, 2001); interval analysis (Alefeld and Herzberger, 1983); p-bounds (Ferson, 2002), and Bayesian methods. Ferson (2002); Ayuub (2001); Ben-Haim (2001); and Burgman (2005) introduce these other methods, while this chapter describes three examples of capturing and using subjective judgement with Bayesian methods. The first captures and uses expert knowledge to infer effects of livestock grazing on bird species. The second represents differences of opinion as different priors and examines how well data leads to agreement. The third represents expert knowledge as a Bayesian network.

However, in using subjective judgement it is important to remember that experts are often wrong and nearly always overconfident (Burgman, 2005). The answers obtained from elicitation exercises will be clouded by the format of the questions, the personal attributes and experiences of the people being asked, and what is being asked. However, the same is also true of the data that we might collect in our scientific endeavours. In collecting data, we aim to minimize biases or at least make the biases consistent so that the results can be calibrated. The same is true when eliciting subjective judgements.

Using the consensus of experts

Martin *et al.* (2005) examined the abundance of birds under three different levels of grazing (low, medium, and high) with eight replicate sites for each grazing level. Twenty experts provided judgements about the relative impacts of grazing on a list of 31 bird species, indicating whether the mean abundance of a species would decline, remain stable or increase under a particular grazing level. The experts only provided responses where they were confident of the effect. The opinions of the experts are a sample of the collective scientific knowledge about the impacts of grazing on birds.

To account for the large number of zeroes in their bird survey data, Martin *et al.* (2005) analysed their data using a two-component model that first predicted whether a species was present at a site and then the abundance of the species conditional on it being present. The first component was based on a logistic regression model while the second was based on a truncated Poisson regression model.

The probability of presence (p_{ij}) of species i at site j was modelled using the logistic regression (Chapter 5) equation:

$$\text{logit}(p_{ij}) = s_{0i} + g_{0j} + t_{0ij},$$

where s_0 is the species effect, g_0 is the site effect and t_{0ij} is the interaction between site and species. These effects were treated as random, and when using uninformative priors (see Chapter 1) the random effects were drawn from normal distributions with a mean of zero and precision that had a gamma(0.1, 0.1) prior.

Similarly, the mean density (λ_{ij}) of each species (i) at each site (j) was modelled with random effects for species, site and the interaction but used a log link function:

$$\log(\lambda_{ij}) = s_{1i} + g_{1j} + t_{1ij}.$$

The responses of the experts provided prior information about the random effects in the above models. The expert responses were coded as −1 (decrease), 0 (stable) or 1 (increase). The mean and variance of these responses was used to construct the prior for the effect of each grazing level and species. The prior was narrower when the experts tended to agree about the impact on a species and was wider when they disagreed. The precision of the prior was equal to the inverse of the variance of the experts' responses. When all experts agreed, the sample variance was equal to zero, making the precision infinity ($1/\infty$). To avoid this numerical difficulty, perfect agreement among all experts was assigned a large precision of 30. The precision was considerably less than 30 in cases where there was at least some disagreement among experts.

Using the means and precisions of the experts' responses as the means and precisions of the priors would imply that the experts' responses and the data were on the same scale, which is not the case. Therefore, the experts' responses were re-scaled by adding a parameter to the mean and multiplying the precision by a second parameter. The required degree of re-scaling was unknown a priori, so uninformative prior distributions were used for these two extra parameters. The parameter for re-scaling the mean could either be positive or negative, so an uninformative normal distribution was used. The parameter for re-scaling the precision must be positive so an uninformative gamma distribution was used.

The WinBUGS code and data for this analysis is provided on the book's website courtesy of Dr Tara Martin. The analysis uses the 'zeroes trick' (similar to the 'ones trick', see Box 7.2 and WinBUGS user manual) to construct the two-component model, accounting for the combination of the logistic regression for presence and the truncate Poisson model for abundance. The rest of the code is straightforward (but long, hence its exclusion from the printed book, see the website) in which the many priors are constructed and the predictions are made for each of the 31 species at the three grazing levels.

A total of 50 000 samples from the posterior distribution after discarding the first 10 000 samples provided the posterior distribution of the mean abundance of the birds under the different levels of grazing. A selection of the results is presented here; further details are provided by Martin *et al.* (2005). For the white-browed scrubwren, the 18 experts who predicted effects of heavy grazing all agreed that it would adversely affect the species. The informative prior improved the precision of the prediction of abundance in this case (Fig. 10.1).

In contrast, the priors for the effects of medium and heavy grazing on rufous songlarks were broad. Therefore, the predictions were driven by the data, with the posterior distributions of abundance being similar for both the informative and uninformative priors (Fig. 10.2).

Experts are subject to a range of biases, so prior information based on only a small sample of experts is likely to lead to biased inference. By eliciting responses from a broad range of experts, the prior information is a sample of the view of the scientific community about the effects of grazing. Martin *et al.* (2005) were able to determine the level of agreement among experts, with close agreement producing precise priors and disagreement producing uninformative priors. In the latter case, the inference was driven largely by the data.

Subjective priors

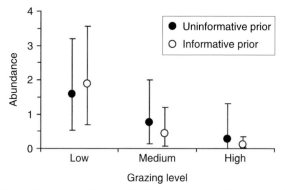

Fig. 10.1 Predicted abundance of white-browed scrubwrens at different levels of grazing, when using uninformative and informative priors (based on models and data by Martin *et al.*, 2005). The dots are the means of the posterior distributions and the bars are the 95% credible intervals.

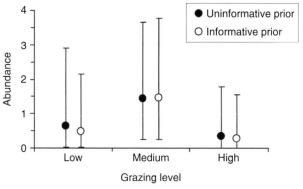

Fig. 10.2 Predicted abundance of rufous songlarks at different levels of grazing, when using uninformative and informative priors (based on models and data by Martin *et al.*, 2005). The dots are the means of the posterior distributions and the bars are the 95% credible intervals.

Representing differences of opinion with subjective priors

Subjective priors are particularly useful when differences of opinion are an important component of applied ecological problems. Many environmental conflicts exist because people disagree about the impacts of a particular activity. For example, while there is little or no disagreement that the Earth is warming, there is some scientific disagreement about the contribution of humans to the warming through our production of carbon dioxide and other greenhouse gases. The differences of opinion could be reflected in different priors for parameters that link increased production of greenhouse gases to the global energy balance.

Crome *et al.* (1996) used Bayesian methods to examine impacts of logging on birds and mammals. Rather than obtaining a consensus among experts as done by Martin *et al.* (2005), Crome *et al.* (1996) were particularly interested in examining actual differences of opinion, which were elicited from 15 interviewees including foresters, conservation activists, and members of the public. A particularly interesting aspect of the work by Crome *et al.* (1996) was examining how effectively results of experimental logging were able to help the different lobby groups reach consensus about the impacts.

Crome *et al.* (1996) measured the capture rate of mammals and birds at 18 rainforest sites. After multiple surveys at each of the sites, nine were logged, and then all 18 sites continued to be surveyed. In the absence of an effect of logging, differences in the capture rate between the logged and unlogged sites prior to logging would be the same as differences after logging. If logging had a negative effect then the capture rate in the logged sites would be reduced relative to the capture rate in the unlogged sites. To help improve normality, the capture rates were log-transformed prior to analysis.

Interviews were used to establish subjective priors for the impacts of logging using a technique developed by Hampton *et al.* (1973). Interviewees were asked to choose levels of impact (expressed as a percentage increase or decrease) that corresponded to different percentiles of the prior distribution. The questions were (Crome *et al.*, 1996):

1. Choose a level of impact so that there is a 50% chance that the effect will be below this level and therefore a 50% chance that the effect will be above this level.
2. Suppose I now tell you that the effect is below this level; tell me your new 50% level: this represents the level at which you think there is a 25% chance that the effect will be below it.
3. Suppose I now tell you that the effect is above your original 50% level; tell me your new 50% level: this represents the level at which you think there is a 75% chance that the effect will be below it.
4. What is the smallest credible level for the effect? By this, I mean what level do you think it is one hundred to one against the effect being below this level?

5. What is the highest credible level for the effect? By this, I mean what level do you think it is one hundred to one against the effect being above this level?
6. Can you be certain that the effect will not be below any level (you would bet your life on it)? If so, what is this level?
7. Can you be certain that the effect will not be above any level (you would bet your life on it)? If so, what is this level?

Crome *et al.* (1996) represented the range of opinions of the 15 interviewees by two polarized points of view (a conservation activist and a forester), and a third point of view that took the middle ground (a lay person):

1. Logging has a strong negative effect (pessimistic prior).
2. Logging has little or no effect (optimistic prior).
3. The impact of logging is related to the amount of canopy removed, which is ~25% reduction in the experimental logging (disinterested prior).

Prior distributions were generated to reflect these three points of view using a parameter (δ) that was the multiplicative effect on the capture rate of the species of birds and mammals in the study. Therefore, $\delta = 1$ represented no change in response to logging, $\delta = 0.75$ represented a 25% decline in capture rate, and $\delta = 1.25$ represented a 25% increase.

The priors were established by fitting a mixed two-component lognormal distribution to the percentile responses (Fig. 10.3). The probability

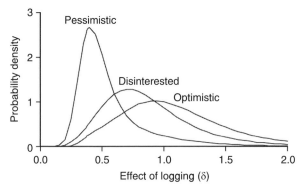

Fig. 10.3 Prior distributions used to represent three opinions about the effect of logging; a strong negative effect (pessimistic), little or no effect (optimistic), and an effect proportional to amount of canopy reduction (disinterested) (re-drawn from Crome *et al.*, 1996.).

density function was equal to:

$$g(x) = wf_1(x) + (1 - w)f_2(x)$$

where $f_1(x)$ and $f_2(x)$ are the probability density functions of regular lognormal distributions and w is a weighing factor that mixes these two distributions in proportion. This distribution was chosen because it permitted only positive values and was sufficiently flexible to accommodate the range of responses.

In addition to inspecting the probability density functions (Fig. 10.3), the priors can be compared by calculating the probability that the logging effect is within particular intervals. For example, the pessimistic prior has 82% of its probability (the area under the probability density function) less than 0.75. Therefore, a priori there is an 82% probability that logging causes at least a 25% reduction in recapture rates. The corresponding probability is 22% for the optimistic prior, and 41% for the disinterested prior. In contrast, the pessimistic prior implies little belief (probability = 4%) that the logging would cause a 25% or greater increase, while the corresponding probability is 31% for the optimistic prior and 15% for the disinterested prior.

The WinBUGS code used to generate the posterior distributions for the effect of logging on grey fantails and white-tailed rats is given below using the pessimistic prior as an example.

```
model
{
  # pess. prior
  f1 ~ dlnorm(-0.848163783, 14.78669534) # 1st log-norm
                                            dist
  f2 ~ dlnorm(-0.405898802, 4.067580628) # 2nd log-norm
                                            dist
  w ~ dbern(0.588969127) prob 1st is used = 0.589,
  2nd is used otherwise
  delta <- w * f1 + (1 - w) * f2 # delta = f1 or f2
  d <- log(delta) # effect of logging on a log scale to
                    conform to data
  pre ~ dnorm(0, 1.0E-6) # diff. in mean ln(capture rate
                            before logging)
  post <- pre + d # diff. in mean ln(capture rate after
                    logging)
    # multiplicative effects are additive on a log scale
```

```
tau ~ dgamma(0.001, 0.001) # precision of capture
                                           rates
for (i in 1:n_pre) # for each pre-logging survey
{
  diff[i] <- logged[i] - unlogged[i] # calculate diff.
                                     between  logged
                                     and     unlogged
                                     sites
  diff[i] ~ dnorm(pre, tau) # assume this diff. is
                             normally distributed
}
for (i in start_post:finish_post) # for each post-
                                     logging survey
{
  diff[i] <- logged[i] - unlogged[i] # calculate diff.
                                     between  logged
                                     and     unlogged
                                     sites
  diff[i] ~ dnorm(post, tau) # assume this diff. is
                               normally distributed
}
}
```

The optimistic prior was used by replacing the above specification of f1, f2 and w by:

```
f1 ~ dlnorm(0.064917859, 7.513236858)
f2 ~ dlnorm(-0.721319814, 13.72305343)
w ~ dbern(0.9269)
```

The disinterested prior was given by:

```
f1 ~ dlnorm(-0.214411583, 4.498261575)
f2 ~ dlnorm(-0.178172325, 10.41757345)
w ~ dbern(0.6750)
```

In the absence of expectations about the effect of logging, an uninformative prior could be used for delta. One possibility would be a lognormal distribution with a low precision (high variance). This would be achieved in the above WinBUGS code by replacing the specification of delta by a lognormal distribution with a precision of 10^{-6}, which would mean the log-transformed values would have a variance of one million.

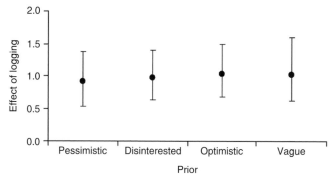

Fig. 10.4 The mean (dot) and 95% credible interval (bars) of the posterior distribution calculated for the effect of logging on white-tailed rats using four different priors.

This large variance permits the possibility that logging has a large positive or negative effect, or the effect is anywhere in between.

The mean and 95% credible intervals of the posterior distributions generated for the data on the white-tailed rat indicate that the priors had only a relatively small influence on the results, with the 95% credible intervals being similar (Fig. 10.4). The three informative priors lead to similar conclusions: that there may be only a small effect of logging because the distributions are centred on a value of 1 representing no multiplicative effect. However, the effect is uncertain for all priors with an increase or decrease of ∼ 50% being possible. The conclusion is qualitatively similar when using an uninformative (vague) prior.

The informative priors have a noticeable effect for the grey fantail data (Fig. 10.5). A large increase in capture rates of grey fantails was observed after the logging, so the vague prior suggests that the effect is likely to be positive and large, with the 95% credible interval spanning the range $1.3 - 20.4$. However, such large increases are very unlikely under the three informative priors, leading to much narrower posterior distributions and upper bounds of the 95% credible intervals ≤ 3.

The posterior probability of δ being less than a particular value can be calculated in WinBUGS. For example, to calculate the probability that $\delta < 0.75$ (corresponding to a decline of 25% or more), the following can be inserted into the above code.

```
P1 <- step(0.75 - delta) # reduction is >25%
```

The step function takes a value of zero when its argument (the term in brackets) is negative and a value of one otherwise. Therefore, the mean of P1 is the probability that delta (δ) is less than 0.75.

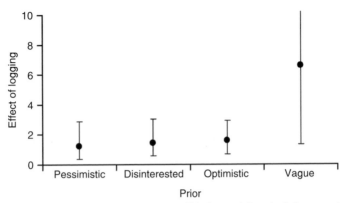

Fig. 10.5 The mean (dot) and 95% credible interval (bars) of the posterior distribution calculated for the effect of logging on grey fantails using four different priors. The upper limit of the 95% credible interval under the vague prior (20.4) is not shown on this graph.

The probability that the capture rate of grey fantail declines by more than 25% in response to logging is small for both the disinterested (10%) and optimistic (3%) priors. However, the relatively vague data mean that the pessimistic prior leads to a much greater posterior probability of decline (31%). While the data led to reasonably good consensus about the effect of logging on white-tailed rats, there is relatively poor consensus about the effect on grey fantails. This result is instructive. In some cases the experimental data can lead people to reach similar conclusions about the effect of logging. In others cases they honestly disagree about the impacts of logging because of their divergent prior (subjective) beliefs and because the data were not sufficiently informative. More data would be required to reach consensus.

Using Bayesian networks to represent expert opinion

While it might be desirable to minimize the use of subjective judgement in science, pressing environmental problems often require that decisions be made with little if any data. In these circumstances, experts are required to make judgements about ecological relationships and parameters. In making these judgements, experts use conceptual models of the system being considered.

For example, we might be interested in the probability that *Pfiesteria*, a toxic alga will be present in a river (Stow and Borsuk,

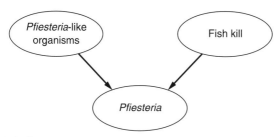

Fig. 10.6 A diagrammatic representation of the occurrence of outbreaks of *Pfiesteria*, a toxic alga believed to occur when fish kills occur in the presence of *Pfiesteria*-like organisms (from Stow and Borsuk, 2003; Burgman, 2005).

2003; Burgman, 2005). One conceptual model for the occurrence of *Pfiesteria* is that it occurs whenever a large number of fish die (a 'fish kill') in the presence *Pfiesteria*-like organisms (PLOs) (Fig. 10.6).

The diagram in Fig. 10.6 is a network that defines the inter-relationships between the components of the system. Each of these components (fish kill, PLOs and *Pfiesteria*) is a node. If the state of each node is treated probabilistically, and the relationships are defined by conditional probability, then the network is known as a Bayesian network. In this example, each node can be in one of two states; the entity is either present or absent. *Pfiesteria* is present whenever fish kills occur in the presence of PLOs, and is absent in all other circumstances.

Therefore, the conditional probability for the occurrence of *Pfiesteria*, given that a fish kill has occurred and PLOs are present is equal to one, and is zero in all other circumstances. Thus:

Pr(*Pfiesteria* | kill and PLO) = 1,
Pr(*Pfiesteria* | kill and no PLO) = 0,
Pr(*Pfiesteria* | no kill and PLO) = 0,
Pr(*Pfiesteria* | no kill and no PLO) = 0.

Based on data for relevant river systems, the independent probability for the occurrence of fish kills is 0.073 and for the occurrence of PLOs is 0.35 (Stow and Borsuk, 2003). Therefore, the probability that PLOs and a fish kill will occur simultaneously equals $0.35 \times 0.073 = 0.026$. Given that *Pfiesteria* outbreaks always occur whenever both PLOs are present and fish kills occur, and under no other circumstances, the probability of *Pfiesteria* outbreaks is simply equal to 0.026.

We now have some numbers to add to the Bayesian network in Fig. 10.6. These describe the probability of being in particular states (Fig. 10.7).

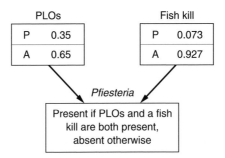

Fig. 10.7 The same model as in Fig. 10.6 but with the probability specified for the presence (P) and absence (A) of PLOs and fish kills.

This simple example of a Bayesian network can illustrate how such networks can be analysed. Software packages other than WinBUGS (e.g. Nettica) are useful for analysing Bayesian networks. However, they can also be analysed in WinBUGS, which also permits continuous states and uncertainty in probability estimates to be included. WinBUGS code for the Bayesian network in Fig. 10.6 is given below:

```
model # (Fig. 8.10b from Burgman, p. 221)
{
  PLO ~ dbern(0.35) # PLOs occur randomly with prob 0.35
  Kill ~ dbern(0.073) # fish kills occur randomly with
                       prob 0.073
  PrPf[1,1] <- 0 # Pr(Pfiesteria outbreak) given PLO and
                   kill absent
  PrPf[1,2] <- 0 # Pr(Pfiesteria outbreak) given PLO
                   absent and kill present
  PrPf[2,1] <- 0 # Pr(Pfiesteria outbreak) given PLO
                   present and kill absent
  PrPf[2,2] <- 1 # Pr(Pfiesteria outbreak) given PLO and
                   kill present
  PLOi <- PLO + 1 # PLO index
  Killi <- Kill + 1 # Kill index
  Pf ~ dbern(PrPf[PLOi,Killi]) # the actual occurrence
                                of a Pfiesteria
                                outbreak
}
```

This model simply treats the occurrence of PLOs and fish kills as Bernoulli events. The probability of *Pfiesteria* being present (PrPf) is equal to one if both PLOs are present and a fish kill occurs, and is zero otherwise. The mean of the WinBUGS samples of Pf estimates the probability of *Pfiesteria* outbreaks. The estimate is 0.025 after 100 000 samples, which is close to the expected value of 0.026. It is easy to calculate this probability without using WinBUGS (0.35×0.073), but Bayesian networks make such calculations easier when there are a large number of dependencies.

This Bayesian network can also be analysed in the opposite direction. For example, given that we do not observe a *Pfiesteria* outbreak (Pf=0), what is the probability that a fish kill has occurred and what is the probability that PLOs are present? By including the data:

```
list(Pf=0)
```

and compiling this with the above model, the posterior distributions for Kill and PLO can be determined. These have a mean of 0.049 and 0.33, which are both less than their independent (prior) probabilities. The absence of *Pfiesteria* means that PLOs and a fish kill are less likely to be present.

The above probabilities can be calculated using Bayes' rule. For example, the prior probability of PLOs being present is 0.35. The probability of observing *Pfiesteria* given that PLOs are present is equal to 0.073 (the probability of a fish kill), so the probability of not observing *Pfiesteria* is 0.927. The other possible hypothesis is that PLOs are not present, which has a prior probability of 0.65 (1−0.035). The probability of not observing *Pfiesteria* in this case is 1. Thus, the posterior probability of PLOs being absent given that *Pfiesteria* is not observed is:

$$\Pr(\text{PLOabsent}) = 0.35 \times 0.927/(0.35 \times 0.927 + 0.65 \times 1.0) = 0.33.$$

Note that the probabilities of PLOs and a fish kill being present both equal one when *Pfiesteria* is observed. This is because both must be present if *Pfiesteria* is observed. This can be analysed in WinBUGS by using the data list(Pf=1) instead of Pf=0.

Bayesian networks help to document the subjective judgements of experts and make the opinions more transparent. These networks can be interrogated with data and compared to other possible representations. For example, Stow and Borsuk (2003) constructed a second model of *Pfiesteria* outbreaks in which PLOs lead to the presence of *Pfiesteria*, which then caused fish kills (Fig. 10.7). Despite being the dominant paradigm at the time, this alternative model had less support from the available data (Stow and Borsuk, 2003).

By analysing complex relationships with Bayesian networks, probabilities are combined in a logical and coherent manner. This is important because humans are bad at estimating probabilities, especially conditional probabilities. For example, few people can determine intuitively the probability of PLOs being present given that *Pfiesteria* is not observed; most people need to do the maths. An analysis with Bayesian networks would be necessary for everyone when analysing more complex networks of relationships.

A final benefit of Bayesian networks is that they can be updated as new data become available. This is especially true if uncertainty in parameters (e.g. probabilities) is treated by representing them as probability distributions. In this case, the probability distributions become the priors that are updated with the addition of data.

Bayesian networks are models; as for all models they are meant to be a simplification. While Bayesian networks are useful for representing and calculating risks using subjective judgement, they encompass only a fraction of the full range of uncertainty. Uncertainty in the probabilities can be considered by treating them as probability distributions. For example, if the standard error for the probability of PLOs being present was 0.1, then we could construct a probability distribution with a mean of 0.35 and standard deviation of 0.1 to represent uncertainty around the estimate.

Beta distributions are useful for describing uncertainty about estimates of probabilities because they are constrained to be between

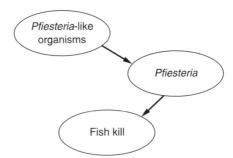

Fig. 10.8 An alternative model for the relationship between *Pfiesteria*-like organisms (PLOs), *Pfiesteria* and fish kills (from Stow and Borsuk, 2003; Burgman, 2005), with fish kills being the result of *Pfiesteria* outbreaks, which in turn occur when PLOs are present.

zero and one. A beta distribution with parameters 7.61 and 14.1 has the required mean and standard deviation for the probability of PLOs being present (see Appendix B for information about determining the parameters of beta distributions). Similarly if the probability of a fish kill had a standard error of 0.05 and mean of 0.073, we could represent uncertainty about its estimate as a beta distribution with parameters 1.90 and 24.2.

Because the probabilities of PLOs being present and a fish kill occurring are uncertain, the probability of a *Pfisteria* outbreak, which is the product of these two values, will also be uncertain (Fig. 10.9).

The density function for the probability of a *Pfisteria* outbreak can be calculated using WinBUGS. The probabilities of PLOs being present and a fish kill occurring are generated using beta distributions and these are then multiplied to give the probability of a *Pfisteria* outbreak, using the following code.

```
model
{
  pPLO ~ dbeta(7.61, 14.1) # beta with mean 0.35 and
                    sd 0.1
  pKill ~ dbeta(1.90, 24.2) # beta with mean 0.073 and
                    sd 0.05
  PrPf <- pPLO*pKill # Pfisteria prob. = pPLO*pKill
}
```

The variable PrPf has a mean of 0.025 and a 95% credible interval of [0.0026, 0.076]. If we can characterize our uncertainty about parameters as probability distributions, we can also propagate uncertainty about those parameters when calculating functions of those parameters such as PrPf. In this example, our 95% credible interval for the probability of a *Pfisteria* outbreak (PrPf) spans more than an order of magnitude.

In calculating the credible interval for a *Pfisteria* outbreak, we assumed that we could represent the two probabilities used in the calculation as probability distributions. However, our subjective judgement may be such that we can only place bounds on the probabilities. For example, we might be sure that the probability of a fish kill is between 0.01 and 0.25 (given the particular hydrological conditions), but unable to specify a probability distribution. Similarly, we may be only able to place bounds of [0.2, 0.5] on the probability of PLOs being present. In this case, the probability of a *Pfisteria* outbreak is also bounded. Assuming that the

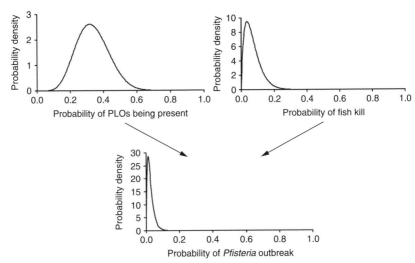

Fig. 10.9 Distributions for the probability of occurrence of PLOs (beta distribution with mean 0.35 and standard deviation 0.1) and fish kills (beta distribution with mean 0.073 and standard deviation 0.05) lead to a distribution for probability of a *Pfisteria* outbreak (mean 0.025 and standard deviation 0.02).

occurrence of PLOs and fish kills are independent events, interval arithmetic provides the bounds (Ferson, 2002; Burgman, 2005). The lower bound is 0.01×0.2 and the upper bound is 0.25×0.5, producing an interval of [0.002, 0.125].

In all the previous calculations it was assumed that the two events (PLOs present and a fish kill) occurred independently. This is not necessarily the case, and dependencies between the probabilities might need to be considered. For example, if PLOs were never present when a fish kill occurred (a perfect negative association), the lower bound on the probability of a *Pfisteria* outbreak would equal $0.01 \times 0.0 = 0.0$. Conversely, if PLOs were always present when fish kills occurred (a perfect positive association), the upper bound would equal $0.25 \times 1.0 = 0.25$. Consideration of the possible dependencies leads to a much wider interval for the probability of a *Pfisteria* outbreak [0.0, 0.25].

The above dependency bound was calculated assuming any form of dependency was possible. Different assumptions about the form of the dependence between the variables (e.g. only permitting positive dependencies) would lead to narrower intervals; for further details see Ferson (2002; 2005) and references therein. The important point is that probabilities in Bayesian networks are usually combined by assuming that events are independent. However, dependencies between events should be considered or outcomes of the calculations may not be reliable.

Concluding remarks

Bayesian methods provide a means of analysing subjective judgement in scientific studies. Bayes' rule permits this subjective judgement to be updated logically and coherently when data become available. The explicit inclusion of subjective judgement in science makes some people uneasy (Fisher, 1930; Dennis, 1996). However, subjectivity can be used beneficially to enhance science. The first example in this chapter demonstrates that experts can be interrogated to represent the current state of knowledge regarding impacts of grazing on birds. The state of knowledge is highly uncertain where the experts disagree, which leads to imprecise priors and the conclusions are dominated by the data. The priors are more precise when they agree and the data have a smaller effect on the conclusions.

Embracing subjectivity is important in many environmental management problems because many environmental conflicts arise owing to differences of opinion rather than differences of fact. In the example by Crome *et al.* (1996), the differences of opinion were reflected in the different priors for the impact of logging on birds and mammals. By collecting and analysing data on logged and unlogged sites, Crome *et al.* (1996) demonstrated that these differences of opinion could reach consensus for many species, with large negative or positive impacts being unlikely regardless of the prior.

The final example in this chapter represented expert opinion as a network of relationships linked by probabilistic rules. Construction and analysis of these Bayesian networks helps to make the subjective judgement that underlies many management decisions more transparent. In these three examples, the value of Bayesian methods is recognizing that subjectivity exists, making it transparent and permitting the subjective judgements to be updated objectively with data.

11

Conclusion

While this book argues that there are clear advantages of using Bayesian methods, it is unlikely that they will be used to the total exclusion of all others. However, there are four important benefits that mean Bayesian methods should be considered by ecologists.

Prior information

Bayesian methods use prior information when available, and the contribution of prior information is transparent. In the absence of prior information, parameter estimates are equivalent to those obtained by maximum likelihood estimation. Those uncomfortable with using prior information often ask: 'What if my prior information is misleading?' Such a question is analogous to: 'What if my data are misleading?' Ecologists do all we can to ensure our data are not misleading. The same needs to apply to prior information.

In the example of measuring the mean diameter of trees in a remnant (Box 3.2), it is unrealistic to assume that we do not have relevant information about the mean diameter of trees when we have measured every tree in 43 other similar remnants. Similarly, when determining whether a species is present at a site or not, it makes sense to use prior information about habitat quality to contribute to the data that are collected (Box 1.5). Prior information can be very cost effective, sometimes representing the equivalent of years of data collection (Box 7.4). So in addition to asking whether the prior is representative, we should also ask whether we can afford to ignore the prior information.

Flexible statistical models

Bayesian statistical models are made to conform to the data, rather than requiring that the data conform to the statistical model. Bayesian methods have great flexibility, so they can easily handle complex analysis, such as those using non-linear or hierarchical random effects models. Other statistical methods can be used for these analyses, but Bayesian methods are easiest (Clark, 2005). Bayesian methods also do not require a collection of different approaches (e.g. null hypothesis testing, likelihood, quasi-likelihood, confidence intervals, permutation tests, etc.) – all Bayesian analyses can be conducted with the one statistical framework based on probability.

Thus, Bayesian non-linear models have the same basic structure as non-linear models (Box 5.6). Hierarchical models can be compared to equivalent non-hierarchical models (Box 3.5). Essentially any functional relationship can be examined using Bayesian methods (Chapter 8), and uncertainty in the resulting parameter estimates can be propagate through to predictions (Box 3.15; Chapter 9).

Intuitive results

The results of Bayesian analyses are intuitive to most ecologists by providing the probability that the parameter estimate is within a particular range, or the probability that the hypothesis is true. Rather than determining, for example, the probability that the data would be obtained if the hypothesis is true, Bayesian methods calculate the probability of the hypothesis being true, which is normally a more useful outcome for ecologists. For example, we can determine the probability that a species is absent from a site when it has not been recorded (Box 1.5), or the probability that a parameter is within a particular range of values by calculating credible intervals (Chapter 1).

Bayesian methods make us think

Bayesian methods require us to think about the appropriate statistical model, and the relevance of our results in relation to previous research. A Bayesian analysis requires answers to questions about the appropriate functional relationships and probability distributions for the statistical models. So rather than using the particular statistical model with its associated assumptions that is contained within our favourite software, we are able to modify the assumptions if necessary.

Perhaps more importantly, Bayesian methods force researchers to consider previous research results explicitly. So rather than obtaining a non-significant result and making the common mistake of concluding that the null hypothesis of no impact is true (Fidler *et al.*, 2006), we might actually conclude that the data are entirely consistent with the impact that was predicted a priori (e.g. the mulgara example in Chapter 6). An explicit consideration of prior information forces us to demonstrate the contribution of the new data to the current state of knowledge. And importantly, we can use multiple lines of evidence to make inference rather than only focusing on the data from the most recent study.

A Bayesian future for ecology

This book has not covered some statistical approaches that are used by ecologists. These include survival analysis, spatial correlation, and multi-variate analysis. Survival analysis and spatial correlation were excluded for the sake of space. There are examples of these in the WinBUGS program and also in the literature (e.g. Wintle and Bardos, in press). Bayesian multi-variate analysis is in its infancy, and is difficult to conduct in WinBUGS. However, the required quantitative routines (e.g. eigen analysis) are being developed in WinBUGS, so Bayesian multi-variate methods are not far off.

The list of topics covered is not exhaustive because this book aims to introduce ecologists to Bayesian methods. As more ecologists use Bayesian methods, the range of examples will grow. Further, useful informative prior distributions for important ecological parameters will also become more available because the posterior distribution for one study can be used as the prior distribution for the next.

Appendices

A

A tutorial for running WinBUGS

Before running WinBUGS, you will need to download it from the WinBUGS website. The latest version of WinBUGS is OpenBUGS, using open source software, which can be obtained at:

http://mathstat.helsinki.fi/openbugs/

Older versions of WinBUGS can be obtained from

http://www.mrc-bsu.cam.ac.uk/bugs/winbugs/contents.shtml

The older versions of WinBUGS require a key for full functionality, which requires that you register (for free). The newer OpenBUGS version does not require a key. Once you have downloaded and installed the software, you are ready to run WinBUGS. The following tutorial is based on the one provided with the program.

A summary of steps for running WinBUGS

1. Select the term 'model' and then 'check model'.
2. Select the data and load.
3. Compile the model.
4. Set initial values for the Markov chain.
5. Set variables to be sampled.
6. Take samples.
7. Check samples.

The steps in more detail

This tutorial steps through how to analyse a mean in WinBUGS, using the example of estimating the average tree diameter in a park (Box 1.8). It is based on the online User Manual in the WinBUGS program.

Run WinBUGS and open a new window (Menu: File/New). Type the following code into the window

```
model
{
  mean ~ dnorm(53, 0.04)      # prior for mean
  prec <- 1/184.9             # precision of the data
  for (i in 1:10)             # for each of the ten trees...
  {
    Y[i] ~ dnorm(mean, prec)  # diameter drawn from a
                                normal distribution
  }
}
list(Y=c(42, 43, 58, 70, 47, 51, 85, 63, 58, 46))
```

The BUGS language allows a concise expression of the model, with the model contained within the curly brackets following the word `model`. The parameter to be estimated is the mean diameter of trees (`mean`). The first line specifies the prior for the mean; in this case it is derived from surveys of other sites in the study area. The second line defines the precision of the data. Precision is the inverse of the variance. In this case, the precision is specified although it could be estimated in addition to the mean. The `for` loop is a concise way of writing that the diameters of each of the ten trees that are in the sample are drawn from the same normal distribution.

The data are listed in the final line, with each number representing one of the measured tree diameters. For example, `Y[1]` is equal to 42, the diameter of the first measured tree.

To obtain samples from the model, first bring down the Model menu, and open the Specification Tool. It is necessary to check that the model description fully defines a probability model. Highlight the key word `model` at the beginning of the model description, say, by double clicking on it with the mouse. Then execute the check model command from the Specification Tool by clicking on the button − 'model is syntactically correct' should appear in the status line at the bottom left corner of the screen.

Next load the data. To check and load the data, highlight the key word list at the start of the data description, and then click the 'load data' button in the Specification Tool − 'data loaded' should appear in the status line.

Next execute the Compile command from the Specification Tool to set up the data structures used by the Gibbs sampler — 'model compiled' should appear in the status line.

Finally the MCMC sampler must be given some initial values. One way to initialize the model is to allow WinBUGS to generate values by drawing samples from the prior distributions. Click 'gen inits' to do this — 'initial values generated: model initialized' should appear in the status line.

A second method is to assign particular initial values to some or all of the stochastic nodes. The only stochastic node in this example is the variable mean. This can be set by using a statement such as

```
list(mean=50).
```

To use an initial value of 50, type this line in the same window as the model, highlight the key word `list` and then execute the 'load inits' command from the Specification Tool — 'model is initialized' should appear in the status line.

Generating initial values randomly works perfectly well in this example. In others, the values that are generated may lead to numerical errors. It is often best to specify reasonable initial values yourself, especially for more complicated statistical models.

Note that you have been led through the various steps needed to check the model, load the data, compile the model, and load initial values. A menu option will be greyed out until it is available.

WinBUGS is now able to generate samples. From the Model menu, open the Update ... Tool, and from the Inference menu open the Samples ... Tool. You may wish to reposition these windows.

Type the name of each node (variable) you wish to monitor in the node box of the Sample Monitor Tool (in this example mean) clicking on 'set' after typing each name. The names are retained in a pull down list. Take 1000 samples of the monitored nodes by clicking 'update' in the Update Tool.

To view traces of all monitored nodes, type * in the node box and click on trace: a Dynamic trace window will open to display the monitored values. Now click on the update button of the Update Tool. The simulated values should be displayed. Simulate through, say, 5000 iterations by changing the number of updates to 5000 and clicking the 'update' button. Then try clicking in turn on stats, history, density, autoC, quantiles buttons of the Sample Monitor Tool, to see summary statistics, traces, kernel density plots, autocorrelation functions, and running quantile estimates. All these plots can be selected by clicking on them, and their size changed by dragging the corners of the bounding box. Clicking on coda generates two files in a format suitable for reading into the program CODA or other software for MCMC convergence diagnostics.

To dispose of some early iterations as a 'burn-in', change the 'beg' (begin) field of the Sample Monitor Tool to, say, 1001 before requesting the analyses. Clicking the stats button of the Sample Monitor Tool will display the statistics of the selected values (nodes) being monitored. The mean or median can be used as the 'best estimate' and the s.d. is equivalent to the standard error in that it measures the uncertainty of the estimate.

For models that include continuous variables (such as mean), the posterior samples are used to construct the density kernels and statistics for the posteriors (e.g. mean and standard deviation). Bayesian confidence intervals (credible intervals) can be constructed by recording the percentiles of the posterior samples. For example, the lower bound of the 95% credible interval for mean is obtained by recording the value below which 2.5% of the samples occur. The upper bound is the value above which 2.5% of the posterior samples occur. These bounds are obtained by clicking the stats button.

There are various ways that the results should be assessed.

1. Do the density plots appear to be truncated? If so, it may indicate that the limits of the prior distribution are having an appreciable influence on the results.
2. Are the density plots smooth or bumpy? Bumpy density plots suggest that too few samples may have been taken. As with any Monte Carlo procedure, the precision of the estimate will increase with more samples. For example, the density kernel based on 1000 samples of mean is bumpy, while 100 000 samples generate a smooth plot.
3. Examine the 'history' of the results. Ideally, the graph will appear like white noise. If there is relatively strong autocorrelation, there may be some discernible pattern, but there should certainly be no suggestion of trend if the samples are to be a reasonable representation of the

posterior distribution. In the tree diameter example, WinBUGS samples directly from the posterior distribution, so the samples have no correlation.
4. Use different initial values to ensure that results are not sensitive to the choice. This can be assessed subjectively, or the Gelman-Rubin convergence statistic can be used. See the WinBUGS online User Manual and associated references for details of the Gelman-Rubin statistic.

Note that in WinBUGS, each variable can only be assigned a value once. With only a small number of specific exceptions, a variable name can only appear once on the left-hand side of the equations in the code. This means that the order of the statements in the WinBUGS code does not matter (e.g. the prior for mean could be specified on the lines before or after it is used in the code). This feature is at first a little odd to users familiar with other programming languages, where the order of statements matters.

How to write WinBUGS code

Writing WinBUGS code gets easier with experience. It is also useful to examine similar models that others have written such as the examples in this book (code available at http://arcue.botony.unimelb.edu.au/bayes.html), online examples in WinBUGS, and those in Congdon (2003). The important attributes of a WinBUGS model are the same attributes of any Bayesian analysis: priors for the parameters being estimated, data, a model that relates the parameters to the data and the posteriors for the parameters. WinBUGS code includes the prior for the parameters, but most of the code is usually the description of how the data are related to the parameters. The posterior is then generated by WinBUGS with Monte Carlo sampling.

When estimating the tree diameter, we are interested in the variable mean. This is the variable that needs to be recorded by WinBUGS. The variable mean is influenced by both the prior, which is specified directly, and through the influence of the model in which the data are drawn from a distribution that uses mean as a parameter. WinBUGS then samples randomly from the posterior distribution of the variable mean by accounting for the prior and the data. All WinBUGS models have a similar format. The priors for the various parameters are specified, and then the code describes a model of how the data were generated. Based on the description of the model, WinBUGS then obtains samples from the posterior by determining the likelihood of obtaining the data given the parameters (the likelihood) and the prior probability of the parameters. Any parameter that is not related to the data in any way will be sampled from its prior distribution – the prior and posterior are the same in the absence of data.

B
Probability distributions

There are many excellent texts on probability and probability distributions. My favourite ones are the collection on all manner of distributions by Johnson, Kotz and others (e.g. Johnson *et al.*, 1992; Johnson *et al.*, 1994; Johnson *et al.*, 1995; Johnson *et al.*, 1997; Kotz *et al.*, 2000; Balakrishnan and Nevzorov, 2003). There is no point trying to replicate these texts here. This appendix describes the probability distributions that are used within this book, plus a few additional distributions that are likely to prove useful for ecologists. There are additional distributions available within WinBUGS, and other distributions can be constructed using the 'ones trick' (see Box 7.2).

The distributions are classified according to whether they are univariate (returning a single random number) or multivariate (returning two or more, possibly related random numbers), and whether they are discrete (returning integers) or continuous (return floating point numbers). The presentation of each distribution concentrates on their possible implementation in Bayesian models. Prior to presenting each distribution, a basic background of discrete and continuous random variables is provided.

Discrete random variables

Discrete random variables are defined by probability distributions that describe the probability of the random variable achieving outcomes. Discrete random variables are the most simple to understand, so we begin with them. The simplest random variable is achieved by tossing a coin, with the outcomes being a head or a tail. For a fair coin, the probability of each outcome is 0.5.

For many random variables, the outcomes can be most easily represented as numbers. In some cases, the identifying numbers are natural, such as when counting the number of individuals of a species, or the number of species. For example, the number of plants seen in a quadrat could be represented as a random variable.

In other cases, the different outcomes are represented by identifying numbers for mathematical convenience. For example, the result of a coin toss could be represented as a zero (representing a head) or one (representing a tail). In this case, the assignation of a head as one is arbitrary; the outcome of a head could just as well be represented as zero. Ecological classes could be assigned an identifying number (class 'zebra' is a one, class 'giraffe' is a two, class 'lion' is a three, etc.). Again, the allocation of outcomes to identifying numbers is arbitrary.

Probability distributions for discrete random variables are usually described by using symbols. For the example of tossing a fair coin, with the outcome given by X, and a head represented by zero and a tail by one, the probability distribution is given as:

$$\Pr(X = x) = 0.5, \text{ for } x = 0 \text{ or } 1.$$

For the 'African mammal example', the probability distribution might be given as:

$$\Pr(X = x) = p_x, \text{ for } x = 0, 1, 2, \ldots N,$$

where N is the number of mammal species being considered.

A requirement of discrete random variables is that the probabilities for all possible outcomes sum to one:

$$\sum_i \Pr(X = i) = 1.$$

The mean of the probability distribution (μ) is equal to:

$$\mu = \sum_i i \times \Pr(X = i),$$

and the variance (σ^2), which measures the spread of the distribution, is equal to:

$$\sigma^2 = \sum_i (i - \mu)^2 \times \Pr(X = i).$$

Other parameters are used to describe the characteristics of a probability distribution. These are based on what are known as the moments of the distribution. The nth central moment (m_n) for discrete probability distributions is given by the expected value of the deviation from the mean raised to the power of n:

$$m_n = \sum_i (i - \mu)^n \times \Pr(X = i).$$

Note that the first central moment is equal to zero, and the second central moment is equal to the variance. Higher order central moments are useful for describing the shape of probability distributions. Two important measures (Balakrishnan and Nevzorov, 2003) include skewness, which measures asymmetry, and kurtosis, which measures the 'fatness' of the tails of a distribution.

Skewness is equal to $m_3/m_2^{3/2}$ and kurtosis is m_4/m_2^2. By using central moments, skewness and kurtosis are measured only by deviations from the mean, not by the mean itself. Further, they are scale-free, since the powers in the numerator (3 and 4, respectively) are balanced by the powers in the denominator ($2 \times 3/2$ and 2×2, respectively).

Positive skewness ($m_3/m_2^{3/2} > 0$) means that the right-hand tail of the probability distribution is longer than the left. Positive skewness (or right skewness) occurs when the mean is greater than the median. The opposite is true for negative skewness.

Kurtosis is often expressed relative to that of a normal distribution (described below). If the distribution has a measure of kurtosis greater than the normal (i.e. >3), then it is referred to as being leptokurtic ('fat-tailed'). The term platykurtic is used to describe 'thin-tailed' distributions for which the measure of kurtosis is less than 3.

Continuous random variables

Continuous random variables can take non-integer (real) numbers rather than being restricted to discrete values. Continuous random variables are defined by a cumulative distribution function $F(x)$ that defines the probability that the random variable (X) is less than some value (x) over some range of real numbers:

$$\Pr(X \leq x) = F(x).$$

$F(x)$ is the cumulative distribution function and by definition, $F(-\infty) = 0$ and $F(\infty) = 1$.

It is possible to determine the probability that a random variable is within the range between x and $x+dx$, where dx is an incremental value:

$$\Pr(x \leq X \leq x + dx) = F(x + dx) - F(x).$$

Calculus provides us with the limit of $\Pr(x \leq X \leq x+dx)$ as dx approaches zero, which is equal to $f(x)dx$, where $f(x) = dF(x)/dx$. The function $f(x)$ is referred to as the probability density function. The actual value of $f(x)$ is not a probability, and can take values greater than one (but not less than zero). However, $f(x)$ does indicate the relative likelihood of different values of the random variable X.

Note that by definition, the cumulative distribution function is:

$$F(x) = \int_{-\infty}^{x} f(u)du.$$

For a continuous random variable with probability density function $f(x)$, the mean (μ) is given by:

$$\mu = \int_{-\infty}^{\infty} xf(x)dx,$$

and the variance (σ^2) is:

$$\sigma^2 = \int_{-\infty}^{\infty} (x - \mu)^2 f(x)dx.$$

The nth central moment is given by:

$$m_n = \int_{-\infty}^{\infty} (x - \mu)^n f(x)dx.$$

Skewness ($m_3/m_2^{3/2}$) and kurtosis (m_4/m_2^2) are measured in the same manner as for discrete distributions.

A range of univariate probability distributions that are used commonly in WinBUGS are summarized in Table A.1. For each distribution, the WinBUGS syntax is provided, along with the parameter/s expressed as a function of the mean and variance of the distribution. Additionally, the mean, variance, skewness and kurtosis are given in terms of the WinBUGS parameters. These formulae permit users to calculate the required WinBUGS parameters if the mean and variance of the distribution is known, or conversely to determine the statistics of the distribution if the WinBUGS parameters are provided. Each of these distributions and a selection of multivariate distributions are described in more detail in the remainder of the appendix.

Table A.1. *A range of univariate probability distributions that are used commonly in WinBUGS. For each distribution, the WinBUGS syntax is provided, along with the WinBUGS parameter/s expressed as a function of the mean (μ) and variance (σ^2) of the distribution. Additionally, the mean, variance, skewness (γ_1) kurtosis (γ_2) are given in terms of the WinBUGS parameters.*

Probability distribution	WinBUGS syntax	Parameters	Mean, variance, skewness and kurtosis
Bernoulli	`dbern(p)`	$p = \mu$	$\mu = p$ $\sigma^2 = p(1-p)$ $\gamma_1 = \dfrac{1 - 2p}{\sqrt{p(1-p)}}$ $\gamma_2 = \dfrac{1}{p} + \dfrac{1}{1-p} - 3$
Binomial	`dbin(p,n)`	$p = 1 - \sigma^2/\mu$ $n = \mu^2/(\mu - \sigma^2)$	$\mu = np$ $\sigma^2 = np(1-p)$ $\gamma_1 = \dfrac{1 - 2p}{\sqrt{np(1-p)}}$ $\gamma_2 = 3 + \dfrac{6p2 - 6p + 1}{np(1-p)}$
Poisson	`dpois(m)`	$m = \sigma^2 = \mu$	$\mu = m$ $\sigma^2 = m$ $\gamma_1 = 1\sqrt{m}$ $\gamma_2 = 3 + 1/m$
Negative Binomial	`dnegbin(p,r)`	$p = \mu/\sigma^2$ $r = \mu/(\sigma^2 - \mu)$	$\mu = r(1-p)/p$ $\sigma^2 = r(1-p)/p^2$ $\gamma_1 = \dfrac{2 - p}{\sqrt{r(1-p)}}$ $\gamma_2 = 3 + \dfrac{6 - p(6-p)}{r(1-p)}$
Categorical	`dcat(p[])`		$\mu = \sum\limits_{i} i \times p[i])$ $\sigma^2 = \sum\limits_{i} (i - \mu)^2 \times p[i]$ $\gamma_1 = \left(\sum\limits_{i} (i - \mu)^3 \times p[i] \right)/\sigma^3$ $\gamma_2 = \left(\sum\limits_{i} (i - \mu)^4 \times p[i] \right)/\sigma^4$

Table A.1. (cont.)

Probability distribution	WinBUGS syntax	Parameters	Mean, variance, skewness and kurtosis
Uniform	`dunif(a,b)`	$a = \mu - \sigma\sqrt{3}$ $b = \mu + \sigma\sqrt{3}$	$\mu = (a+b)/2$ $\sigma^2 = (b-a)^2/12$ $\gamma_1 = 0$ $\gamma_2 = 9/5$
Beta	`dbeta(a,b)`	$a = \mu(\mu(1-\mu)/\sigma^2 - 1)$ $b = (1-\mu)(\mu(1-\mu)/\sigma^2 - 1)$	$\mu = a/(a+b)$ $\sigma^2 = \dfrac{ab}{(a+b)^2(a+b+1)}$ $\gamma_1 = \dfrac{2(b-a)\sqrt{1+a+b}}{\sqrt{ab}(2+a+b)}$ $\gamma_2 = 3 + \dfrac{\left(\begin{array}{c}6[a^3 + a^2(1-2b) \\ +b^2(1+b) \\ -2ab(2+b)]\end{array}\right)}{ab(2+a+b)(3+a+b)}$
Normal	`dnorm(m,t)`	$m = \mu$ $t = 1/\sigma^2$	$\mu = m$ $\sigma^2 = 1/t$ $\gamma_1 = 0$ $\gamma_2 = 3$
Student-t	`dt(m,t,k)`	$m = \mu$ $t = (2\gamma_2 3)/(\gamma_2 \sigma^2)$ $k = 4 + 6/(\gamma_2 - 3)$	$\mu = m$ $\sigma^2 = (k/(k-2))/t$ $\gamma_1 = 0$ $\gamma_2 = 3 + 6/(k-4)$
Lognormal	`dlnorm(a,t)`	$a = \ln\mu - 0.5\ln c$ $t = 1/\ln c$ $c = \sigma^2/\mu^2 + 1.$	$\mu = e^{a+1/2t}$ $\sigma^2 = e^{2a+1/t}(e^{1/t} - 1)$ $\gamma_1 = \left(\sqrt{e^{1/t}} - 1\right)\left(2 + e^{1/t}\right)$ $\gamma_2 = e^{4/t} + 2e^{3/t} + 3e^{2/t} - 3$
Exponential	`dexp(m)`	$m = 1/\mu = 1/\sigma$	$\mu = 1/m$ $\sigma^2 = 1/m^2$ $\gamma_1 = 2$ $\gamma_2 = 9$
Weibull	`dweib(v,m)`	Closed-form solutions for v and m with respect to μ and σ^2 are not available.	$\mu = m^{-1/v}\Gamma(1 + 1/v))$ $\sigma^2 = \left\{\Gamma\left(\dfrac{v+2}{2}\right)\right.$ $\left. - \left[\Gamma\left(\dfrac{v+2}{2}\right)\right]^2\right\}m^{-2/v}$ See description of the Weibull distribution for skewness and kurtosis.

Table A.1. (cont.)

Probability distribution	WinBUGS syntax	Parameters	Mean, variance, skewness and kurtosis
Gamma	dgamma(r,n)	$r = \mu^2/\sigma^2$ $n = \mu/\sigma^2$	$\mu = r/n$ $\sigma^2 = r/n^2$ $\gamma_1 = 2/\sqrt{r}$ $\gamma_2 = 3 + 6/r$
Chi-squared	dchisqr(k)	$k = \mu = \sigma^2/2$	$\mu = k$ $\sigma^2 = 2k$ $\gamma_1 = 2\sqrt{(2/k)}$ $\gamma_2 = 3 + 12/k$

Univariate discrete distributions

Bernoulli

The Bernoulli distribution is defined for two possible outcomes ($X = 0$ or 1). The probability that $X = 1$ is equal to p and the probability that $X = 0$ is equal to $1 - p = q$:

$$\Pr(X = 1) = p,$$
$$\Pr(X = 0) = 1 - p.$$

The mean (μ) of a Bernoulli random variable is equal to p and the variance (σ^2) is equal to $p(1 - p)$.

The Bernoulli distribution can be used to represent the outcome of single random events or trials, with a value of one indicating that the event occurred (the trial succeeded) and a value of zero indicating that the event did not occur (or the trial failed).

In WinBUGS, the Bernoulli distribution is expressed as:

```
dbern(p),
```

where p is the probability of the event occurring, and it returns a value of one (the event occurred) or zero.

Binomial

The binomial distribution is generated as the sum of n independent and identically distributed Bernoulli random variables. It can be viewed as the number of successes out of n independent Bernoulli trials that each have a probability of success equal to p. For the binomial distribution:

$$\Pr(X = x) = \frac{n!}{(n - x)!x!}p^x(1 - p), \quad x = 0, 1, 2, \ldots, n,$$

where x is the number of 'successes' and $n-x$ is the number of 'failures'.

The mean (μ) is equal to np and the variance (σ^2) is equal to $np(1 - p)$.

The binomial distribution is a useful model for describing, for example, the number of survivors or number of encounters of a finite number of individuals or species where the chance of survival or encounter of those individuals or species can be considered identical and the actual outcomes for each individual are independent.

In most circumstances, the value of n is known, while the probability of 'success' p is uncertain. Consider the case in which p has a beta distribution for the prior with parameters a and b (see the beta distribution in the next section), and we collect data for which there are x 'successes' and $n-x$ 'failures'. In this case, the posterior distribution for p under a binomial model will also have a beta distribution, but with parameters $a+x$ and $b+n-x$. Conjugacy occurs in such circumstances when the prior and posterior distributions have the same distributional form (see later section on conjugacy). Thus, the beta distribution is conjugate to the binomial model of observations.

In WinBUGS, the binomial distribution is expressed as:

```
dbin(p, n),
```

where p is the probability of success and n is the number of trials. It returns integers between zero and n.

Poisson

Consider the case in which the parameter n of the binomial model increases towards infinity while at the same time the parameter p decreases towards zero such that the product of the two is equal to a constant ($\lambda = np$). Under this model, there are an enormous number of events that could occur, but each one has a small chance such that on average only λ will occur. In this case, the limit of the binomial distribution as n approaches infinity subject to the constraint $\lambda = np$, is given by the Poisson distribution:

$$\Pr(X = x) = e^{-\lambda}\lambda^x/x!, \quad x = 0, 1, 2, \ldots$$

Both the mean (μ) and variance (σ^2) of the Poisson distribution are equal to λ.

The Poisson distribution is a useful model of counts, because it defines the distribution of the number of events within a time or space when the events are distributed entirely at random (hence the term 'Poisson process' for defining randomly occurring events).

The conjugate distribution for the Poisson model is the gamma distribution. If the prior for the average density of events (λ) has a gamma distribution with parameters a and b, and there is a total of x observations from a sample size of n, then the posterior distribution for λ will have a gamma distribution with parameters $a+x$ and $b+n$.

In WinBUGS, the Poisson distribution is expressed as:

```
dpois(m),
```

where m is the parameter of the Poisson distribution that is equal to both the mean and variance. It returns non-negative integer values (0, 1, 2, 3, ...).

Negative binomial

If the outcomes of Bernoulli events are occurring randomly, then we could count the number of failures until a certain number of successes. Under such a model, the number of failures would have a negative binomial distribution.

Under the negative binomial distribution:

$$\Pr(X = x) = \frac{(x + r - 1)!}{x!(r - 1)!} p^r (1 - p)^r, \quad x = 0, 1, 2, \ldots$$

where p is the probability of success and r is the number of successes before x failures.

The mean (μ) is equal to $r(1-p)/p$ and the variance (σ^2) is equal to $r(1-p)/p^2$.

If $r = 1$, then we are modelling the number of failures until the first success. This special case of the negative binomial is known as the geometric distribution.

In WinBUGS, the negative binomial distribution is expressed as:

```
dnegbin(p, r),
```

where p is the probability of success p and r is the number of successes r.

The negative binomial distribution is also useful when modelling counts that are clumped. While the Poisson distribution is useful for counts when events or individuals are distributed at random, clumping will occur in many circumstances. For example, if examining the density of individuals, there are likely to be some areas or time periods that are likely to have higher densities that others. In these circumstances, the variance among counts would be greater than assumed under the Poisson distribution. If it is assumed that the average density varies as a gamma distribution (see continuous distributions below) among replicates while the actual count of a replicate is drawn from a Poisson distribution, then counts will follow a negative binomial distribution. This is a more general form of the negative binomial in which the parameter r can take non-integer positive values.[1] This is most easily represented in WinBUGS by modelling the mixture of the gamma and Poisson explicitly as a hierarchical model (see Boxes 3.5 and 3.6).

Categorical

The categorical distribution can be used to define an arbitrary probability distribution with a finite number of classes. The probability distribution is defined by specifying the probability that the replicate is obtained from class 1 to k, where k is the number of possible classes. The only constraint is the probabilities sum to one (i.e. it is a probability distribution):

$$\sum_{i=1}^{k} \Pr(X = i) = 1.$$

In WinBUGS, the first class is indexed by the number one, so values of zero cannot be returned.

In WinBUGS, the categorical distribution is expressed as:

`dcat(p[]),`

where `p[]` is a vector of k probabilities that sum to one, and `p[i]` is the probability that the random variable takes a value of `i` ($i = 1, 2, \ldots, k$).

Because categorical distributions in WinBUGS can return only positive integers, it may be necessary to transform the data to conform to the distribution. For example, if Bernoulli successes in the data are coded as

---

[1] Negative binomial distributions in which the parameter r is an integer (as implemented in WinBUGS) are often referred to as Pascal distributions to distinguish them from the more general form of the negative binomial.

ones and failures as zero, an implementation of the Bernoulli distribution with WinBUGS could be coded as:

```
model
{
  p[1] ~ dunif(0, 1)
  p[2] <- 1 - p[1]
  Y1 <- Y+1
  Y1 <- dcat(p[])
}
```

This code is the equivalent of:

```
model
{
  p ~ dunif(0, 1)
  Y <- dbern(p[])
}
```

In this code, the single piece of data is represented by the variable Y, which would have a value of zero or one.

As a second example of dcat(), we can formulate the analysis of a Poisson model. For example, when analysing the helmeted honeyeater productivity data (Box 4.1), it is necessary to calculate the probabilities of there being 0, 1, 2, 3, 4 or 5 offspring. The Poisson model can return values up to infinity, so this can be accommodated in WinBUGS by pooling all values for the random variable that are greater than the largest observed value. Additionally, because the smallest index for dcat() is one, it is necessary to represent the probability of there being x offspring as p[x+1].

The code for the model would be:

```
model
{
  lambda ~ dgamma(0.001, 0.001)
  for (i in 1:6)   # generate probabilities for 0 to 5
  {
    x[i] <- i-1
    p[i] <- exp(-lambda + x[i]*log(lambda) -
    logfact(x[i]))
    # p = exp(-m)*m^x/x! = exp(-m + x*ln(m) - logfact(x))
  }
```

```
p[7] <- 1 - sum(p[1:6])    # this is the probability of 6
                             or more offspring
for (i in 1:35)
{
  OClass[i] <- Offspring[i] + 1
# class id needs to be incremented by 1 to accommodate
  zero offspring
  OClass[i] ~ dcat(p[1:7])
}
}
```

This code generates the same posterior distribution for lambda as that used in Box 4.1.

Univariate continuous distributions

Uniform

The standard uniform distribution is the simplest continuous distribution, but is particularly useful as an uninformative prior distribution for proportions. The probability density function is defined as $f(x) = 1$ for values of x between zero and one, and $f(x) = 0$ otherwise. The cumulative distribution function on the interval $0 \leq x \leq 1$ is $F(x) = x$.

The uniform distribution can be generalized from the interval $[0, 1]$ to any interval $[a, b]$, where $b > a$. In this case, the probability density function is $f(x) = 1/(b-a)$ for values of x between a and b. The cumulative distribution function on the interval $a \leq x \leq b$ is $F(x) = (x-a)/(b-a)$.

This generalized form of the uniform distribution has a mean (μ) of $(a+b)/2$ and variance (σ^2) of $(b-a)^2/12$.

In particular, the standard uniform distribution ($a = 0$ and $b = 1$) has a mean (μ) of 0.5 and variance (σ^2) of $1/12$.

In WinBUGS, the uniform distribution is expressed as:

```
dunif(a,b),
```

where a is the lower limit and b is the upper limit.

Beta

The beta distribution is useful for modelling proportions as it is the conjugate distribution for binomial sampling. If p has a beta distribution for the prior with parameters (a and b), the data are x 'successes' and f 'failures', then the posterior distribution for p under a binomial model will also have a beta distribution, but with parameters $a+x$ and $b+f$.

The standard uniform distribution is a special case of the beta distribution, for which $a=b=1$. If there are x successes and f failures from $x+f$ trials, and if the prior is an 'uninformative' uniform distribution then the posterior will be a beta distribution with parameters $x+1$ and $f+1$. Therefore, the parameter a can be thought of as the 'number of previous successes plus one' and b as the 'number of previous failures plus one'.

The beta distribution is defined on the interval [0, 1]. The probability density function is:

$$f(x) = \frac{1}{B(a,b)} x^{a-1}(1-x)^{b-1},$$

where $B(a,b)$ is the beta function, and $a>0$ and $b>0$. The beta function is expressed as an integral:

$$B(a,b) = \int_0^1 t^{a-1}(1-t)^{b-1}dt = \frac{\Gamma(a)\Gamma(b)}{\Gamma(a+b)}.$$

For integer values of x, $\Gamma(x)=(x-1)!$, but more generally for any value of x:

$$\Gamma(x) = \int_0^\infty t^{x-1}e^{-t}dt.$$

The expression for the beta function that uses the gamma function ($\Gamma()$) is useful because the gamma function (or $\ln(\Gamma())$ is more commonly available in quantitative packages (e.g. Microsoft Excel has the function GAMMALN()) than the beta function.

The beta distribution has a mean (μ) of $a/(a+b)$ and variance (σ^2) of $ab/(a+b)^2(a+b+1)$.

Given a particular mean (μ) and variance (σ^2), it is possible to determine the appropriate parameters (a and b) for the beta distribution:

$$a = \mu[\mu(1-\mu)/\sigma^2 - 1]$$
$$b = (1-\mu)[\mu(1-\mu)/\sigma^2 - 1]$$

Note that if the random variable X has a beta distribution with parameters a and b, then $1-X$ has a beta distribution with parameters b and a (the two distributions are simply reflections of each other).

In WinBUGS, the beta distribution is expressed as:

```
dbeta(a,b),
```

where a and b are the parameters *a* and *b* of the beta distribution.

Normal

The sum of random numbers will approach a normal distribution as the number of random numbers in the sum approaches infinity. This is a particularly useful property because, for example, the mean is simply the sum of random numbers that has been rescaled by dividing by the sample size. The rate at which a sum of random variables converges towards a normal distribution depends on how closely the original random variables resemble a normal. If the original random variables are normal, then the sum will be a normal. However, if the original random variables are, for example, highly skewed, then the sum will only approximate a normal when a sufficient number of variables are summed.

The probability density function of a normal distribution is:

$$f(x) = \sqrt{\frac{\tau}{2\pi}} \exp\left(-\frac{(x-a)^2\tau}{2}\right),$$

with the location parameter *a* and scale parameter $\tau > 0$. It turns out the mean (μ) of the normal distribution is equal to the location parameter (a) and the variance (σ^2) is equal to $1/\tau$. The parameter τ can be referred to as the precision, with larger values of τ leading to smaller variances.

The normal distribution is conjugate to the normal model. If the prior for the mean has a normal distribution (with mean *m* and precision *p*), and the data are drawn form a normal distribution (such that the likelihood has mean *n* and precision *q*), then the posterior for the mean has a normal distribution (with mean $(m/q+n/p)/(1/p+1/q)$ and precision $p+q$).

In WinBUGS, the normal distribution is expressed as:

```
dnorm(mean, tau),
```

where mean is equal to μ and tau is equal to the precision τ ($=1/\sigma^2$).

Student-*t*

If a sample of size n is obtained from a normal distribution, then the quantity:

$$t = \frac{\bar{x} - \mu}{s/\sqrt{n}},$$

will follow Student's standard *t*-distribution, where $\bar{x}$ is the sample mean, s is the sample standard deviation and μ is the population mean. The standard *t*-distribution can be generalized by including a location (m) and scale parameter $(1/\sqrt{t})$ that change the mean and variance respectively.

The probability density function is:

$$f(x) = \frac{\Gamma(\frac{k+1}{2})}{\Gamma(\frac{k}{2})} \sqrt{\frac{t}{k\pi}} [1 + \frac{t}{k}(x - m)^2]^{-(k+1)/2},$$

where $k = n-1$ (the degrees of freedom), which takes values ≥ 2.

The mean of the *t*-distribution is equal to m and the variance is equal to $k/[t(k-2)]$. The *t*-distribution approaches a normal distribution with mean m and variance $1/t$ as k increases towards infinity. The main difference between the two distributions is that the *t*-distribution has 'fatter tails' (greater kurtosis) than the normal distribution.

The *t*-distribution is expressed in WinBUGS as:

```
dt(m, t, k)
```

where m $= m$, t $= t$ and k $= k$.

Lognormal

Just as the normal distribution is the limiting distribution for the sum of random variables, the lognormal distribution is the limiting distribution for the product of positive random variables. A lognormal distribution (Y) is produced on transforming a normal distribution (X) with the exponential function, and a normal distribution is produced when taking the natural logarithm of a lognormal distribution:

$$Y \sim \exp(X), \text{ and}$$
$$\ln(Y) \sim X.$$

The probability density function of the lognormal distribution is:

$$f(x) = \sqrt{\frac{\tau}{2\pi}} \frac{1}{x} \exp\left(-\frac{(\ln x - a)^2 \tau}{2}\right),$$

where a and τ are the mean and precision respectively of the corresponding normal distribution. The mean of the lognormal distribution (μ) is equal to $e^a \sqrt{w}$ and the variance (σ^2) is equal to $e^{2a} w(w-1)$, where $w = e^{1/\tau}$.

Given the mean and variance of the lognormal distribution (μ and σ^2), it is possible to calculate the parameters a and τ:

$$a = \ln \mu - 0.5 \ln c$$
$$\tau = 1/\ln c,$$

where $c = \sigma^2/\mu^2 + 1$.

In WinBUGS, the lognormal distribution is expressed as:

```
dlnorm(mean, tau),
```

where `mean` is equal to a and `tau` is equal to the precision τ of the corresponding normal distribution.

Exponential

Given that events occur completely at random throughout time or along a transect, the time or distance between events will follow an exponential distribution. Because the rate at which events occur is constant with the time or distance since the last event, the time or distance from a random point until the next event also follows an exponential distribution.

The probability density function for the exponential distribution is:

$$f(x) = \lambda \exp(-\lambda x), \quad \text{for } x \geq 0, \text{ and zero otherwise,}$$

where λ is the average density of events.

The cumulative distribution function is:

$$F(x) = 1 - \exp(-\lambda x), \quad \text{for } x \geq 0, \text{ and zero otherwise.}$$

The mean of the exponential distribution is equal to $1/\lambda$, and the variance is equal to $1/\lambda^2$.

In WinBUGS, the exponential distribution is expressed as:

```
dexp(lambda),
```

where `lambda` is equal to λ.

Weibull

Under the exponential model, the rate at which events occur does not change with the time since the event. The Weibull model provides a generalization such that the rate at which events occur (h) is a power function of time since the last event ($h = \lambda v x^{v-1}$). The probability density function in this case is given as:

$$f(x) = \lambda v x^{v-1} \exp(-\lambda x^v), \quad \text{for } x \geq 0, \text{ and zero otherwise.}$$

The cumulative distribution function is:

$$F(x) = 1 - \exp(-\lambda x^v), \quad \text{for } x \geq 0, \text{ and zero otherwise.}$$

The rate at which events occur increases with time or distance since the last event when $v > 1$, and decreases when $v < 1$. The exponential model is a special case of the Weibull with $v=1$.

The mean and variance of the Weibull distribution are expressed in terms of the gamma function:

$$\mu = \lambda^{-1/v}\Gamma(1/v + 1), \text{ and } \sigma^2 = \lambda^{-2/v}[\Gamma(2/v + 1) - \Gamma(1/v + 1)^2].$$

The skewness is equal to:

$$\gamma_1 = 3 + \frac{2\Gamma^3(1 + 1/v) - 3\Gamma(1 + 1/v)\Gamma(1 + 2/v) + \Gamma(1 + 3/v)}{[\Gamma(1 + 2/v) - \Gamma^2(1 + 1/v)]^{3/2}}$$

And the kurtosis is equal to:

$$\gamma_2 = \frac{\left(\begin{array}{c} -6\Gamma^4(1 + 1/v) + 12\Gamma^2(1 + 1/v)\Gamma(1 + 2/v) - \\ 3\Gamma^2(1 + 2/v) - 4\Gamma(1 + 1/v)\Gamma(1 + 3/v) + \Gamma(1 + 4/v) \end{array} \right)}{[\Gamma(1 + 2/v) - \Gamma^2(1 + 1/v)]^2}$$

In WinBUGS, the Weibull distribution is expressed as:

`dweib(v,lambda),`

where `lambda` is equal to λ and `v` is equal to v.

Gamma

The gamma distribution plays important and varied roles in Bayesian statistics. It is the conjugate of the parameter of the Poisson model, it is the conjugate for the precision of the normal model, and if the parameter of the Poisson distribution is mixed by a gamma distribution, the result is a negative binomial distribution.

The gamma distribution is defined for positive real numbers and has the probability density function:

$$f(x) = \frac{v^r x^{r-1} e^{-vx}}{\Gamma(r)}; \quad x > 0.$$

The mean (μ) is equal to r/v and the variance (σ^2) is equal to r/v^2. Thus, given the mean and variance of a gamma distribution, it is straightforward to determine the appropriate values for the parameters r and v: $r = \mu^2/\sigma^2$ and $v = \mu/\sigma^2$.

In WinBUGS, the gamma distribution is expressed as:

```
dgamma(r,nu),
```

where r is equal to r and nu is equal to v.

Chi-squared

The special case of the gamma distribution in which $v = 1/2$ and $r = k/2$, where k is a positive integer, is known as the chi-squared distribution with k degrees of freedom.

In WinBUGS, the chi-squared distribution is expressed as:

```
dchisqr(k),
```

where k is equal to k.

Multivariate discrete distributions

Multinomial

The binomial distribution provides an n sample generalization of the Bernoulli distribution. Similarly, the multinomial distribution provides an n sample generalization of the categorical distribution. Consider N independent and identical trials that each can result in one of k different events or classes. If x_i is the number of the N trials that result in event or class i, then the probability of obtaining the vector of outcomes $(x_1, x_2, \ldots, x_k)$ is equal to:

$$\Pr(X_1 = x_1, X_2 = x_2, \ldots, X_k = x_k) = \frac{N}{\prod\limits_{i=1}^{k} x_i!} \prod\limits_{i=1}^{k} p_i^{x_i}; \quad 0 < p_i < 1 \text{ and } \sum\limits_{i=1}^{n} p_i = 1,$$

where p_i is the probability of a trial resulting in event or class i.

The mean number of trials resulting in event i is equal to Np_i and the variance is equal to $Np_i(1-p_i)$.

The uniform distribution is often used as an uninformative prior for the probability of the binomial model. The uniform distribution is a special case of the beta distribution with both its parameters equal to one, and the beta distribution is the conjugate of the binomial. Just as the multinomial model is the multivariate analogue of the binomial, the Dirichlet is the multivariate analogue of the beta distribution. Thus, the Dirichlet is the conjugate for the probabilities of the multinomial model and an uninformative Dirichlet distribution is commonly expressed by setting its parameters equal to one.

In WinBUGS, the multinomial distribution is expressed as:

```
dmulti(p[], N),
```

where p[] is a vector of k probabilities that sum to one, p[i] is the probability that the random variable takes a value of i ($i = 1, 2, ..., k$), and N is the total number of trials (N).

Multivariate continuous distributions

Multivariate normal

The multivariate normal distribution permits generation of correlated normal random numbers. The probability density function is:

$$f(x_1, x_2, \ldots, x_d) = (2\pi)^{-d/2}|T|^{1/2}\exp[-Q(x_1, x_2, \ldots, x_d)/2],$$

where $Q(t) = tTt'$, $t = (x_1-\mu_1, x_2-\mu_2, \ldots, x_d-\mu_d)$ is a vector of random numbers from which their mean is subtracted, T is the inverse of the variance-covariance matrix, $|T|$ is its determinant and d is the number of correlated normal random numbers being returned.

In WinBUGS, the multivariate normal distribution is expressed as:

```
dmnorm(mu[], T[,]),
```

where mu [] is a vector that contains the means of the normal random variates ($\mu_1, \mu_2, \ldots, \mu_d$) and T[,] is the inverse of the variance-covariance matrix of the variates (T).

Dirichlet

Just as the beta distribution is the conjugate of the probabilities of the binomial model, the Dirichlet distribution is the conjugate of the probabilities of the multinomial model.

The probability density of the Dirichlet distribution is:

$$f(p) = \frac{\Gamma(\sum\limits_{i=1}^{n} \alpha_i)}{\prod\limits_{i=1}^{n} \Gamma(\alpha_i)} \prod_{i=1}^{n} p_i^{\alpha_i - 1}; \quad \text{for } 0 < p_i < 1 \text{ and } \sum_{i=1}^{n} p_i = 1.$$

The parameters α_i determine the probability of obtaining the n random variables p_i. The beta distribution is a special case of the Dirichlet distribution, with $n=2$, $\alpha_1=a$ and $\alpha_2=b$. Just as the parameters a and b in the beta distribution can be considered as the 'number of previous successes and failures plus one' in binomial sampling, each of the α_i parameters can be considered as the 'number of previous occurrences of outcome i plus one' in multinomial sampling.

The mean of the variable p_i is equal to α_i/A where $A = \sum\limits_{j=1}^{n} \alpha_j$ and the variance is equal to $\alpha_i(A - \alpha_i)/A^2(A + 1)$.

In WinBUGS, the Dirichlet distribution is expressed as:

```
ddirch(alpha[]),
```

where `alpha[]` is a vector containing the parameters α_i (α_1, α_2, ..., α_n).

Wishart

The Wishart distribution is the multivariate analogue of the gamma distribution, and therefore the chi-squared distribution. In the same way that the gamma distribution is the conjugate prior for the inverse of the variance for a normal model, the Wishart distribution is the conjugate for the inverse of the variance-covariance matrix of a multivariate normal model. The probability density function is relatively complex and its equation provides little insight (Bilodeau and Brenner, 1999). Its important features are represented in the WinBUGS expression of the distribution:

```
dwish(R[,], k),
```

where R[,] is a square matrix with the same dimensions as the variance-covariance matrix ($p \times p$), and k is the degrees of freedom of the matrix. The matrix R/k represents an estimate of the variance-covariance matrix, with the estimate based on k degrees of freedom (check by simulation).

An uninformative Wishart distributions can be specified by using as small a value of k as possible (for a variance-covariance matrix, $k \geq p$), and where the elements of R represent an assessment of the order of magnitude of the variance-covariance matrix (WinBUGS manual).

Conjugacy

We know that the mean density of trees could conceivably be any positive value. The lognormal distribution is useful as a prior distribution for the mean because it is constrained to be greater than zero, but is unbounded (Box 3.4). I chose the lognormal distribution in this example because it is familiar to ecologists, but any continuous distribution that is constrained to positive values could have been used. For example, a uniform distribution between zero and some large number would also be suitable. If the upper limit on this uniform prior distribution is assumed to be infinity, then the posterior will have a gamma distribution (Hilborn and Mangel, 1997). Additionally, if the prior has a gamma distribution, and we assume the data are drawn from a Poisson distribution, then the posterior will again have a gamma distribution. Therefore, an uninformative gamma distribution would have been another possible choice.

When the prior and the posterior have the same distributional form (although different parameters) for a given model, as in the case of the gamma and the Poisson, then the distribution is said to be conjugate to the model. I introduced a conjugate pair in Chapter 1, where the normal distribution for the mean is the conjugate for the normal model. There are numerous other conjugate pairs. For example, the gamma distribution for the precision parameter is the conjugate for the normal model. Therefore, the gamma distribution occurs quite commonly in Bayesian statistics.

An uninformative gamma distribution can be specified instead of a lognormal distribution by replacing the prior for the mean in Box 3.4 by the following line of code:

```
m ~ dgamma(0.001, 0.001)  # gamma prior for mean trees
                                per quadrat
```

This prior has a mean of 1 but a large variance (1000), so it can take a broad range of values. This uninformative prior leads to the same credible interval as the lognormal prior. Therefore, the choice between these two priors is not particularly important. Choosing conjugate priors can help the WinBUGS algorithms operate more efficiently when sampling from the posteriors.

C
MCMC algorithms

The Metropolis algorithm (Metropolis *et al.*, 1953) is one of the simplest MCMC algorithms used for Bayesian analysis. The following describes how the Metropolis algorithm obtains samples from a posterior distribution. An example of sampling from the posterior distribution of a single parameter is described, but MCMC methods are easily extended to consider multiple parameters simultaneously (Gilks *et al.*, 1996).

The Metropolis algorithm works by starting with an initial arbitrary value for the parameter X_0, which is the first value of the Markov chain. We are interested in obtaining subsequent values of X_t such that they are samples of a random variable with probability density function $p(\ )$.

A new possible value (Y) is generated by drawing it from an arbitrary symmetric probability distribution. This proposal distribution is defined by its probability density function; given the current value X_t, the probability of drawing the value of Y as the possible next value of the Markov chain is equal to $q(Y \mid X_t)$.

Next, the acceptance probability is calculated, which is

$$R(Y|X_t) = \min[1, p(Y)/p(X_t)].$$

Therefore, if the ratio of the posterior probabilities is greater than or equal to 1 (i.e. $p(Y) \geq p(X_t)$), then Y is chosen as the next value of the Markov chain ($X_{t+1} = Y$). If on the other hand $p(Y) < p(X_t)$, then Y is chosen as the next value of the Markov chain with probability $p(Y)/p(X_t)$, and X_t is chosen otherwise.

The important point to note is that the algorithm depends on the ratio of the probability density function at two different points ($p(Y)$ and $p(X_t)$). Based on Bayes' rule, these two values are equal to:

$$p(Y) = \frac{f(Y)\ell(Y)}{\int_{-\infty}^{\infty} f(x)\ell(x)dx}, \text{ and}$$

$$p(X_t) = \frac{f(X_t)\ell(X_t)}{\int_{-\infty}^{\infty} f(x)\ell(x)dx},$$

where $f()$ is the prior probability density function and $\ell()$ is the likelihood function. Because both expressions have the same denominator (i.e. the same scaling constant), the ratio of the two values is simply equal to the ratio of the prior probabilities and likelihoods; the integral is not calculated.

Hastings (1970) modified the Metropolis algorithm to permit non-symmetric distributions to be used for generating the new possible value. For the Metropolis-Hastings algorithm the acceptance probability is equal to $\min[1, \; p(Y)q(X_t \mid Y)/p(X_t)q(Y \mid X_t)]$. This reduces to the original Metropolis algorithm when the proposal distribution is symmetric, $q(Y \mid X_t) = q(X_t \mid Y)$. The Gibbs algorithm (after which WinBUGS is named) is a special case in which $q()$ is chosen such that R is always equal to 1.

The Metropolis algorithm is illustrated below using results of three successes from 12 independent Bernoulli trials. Let s be the probability of success. The probability of observing the data (i.e. the likelihood) is proportional to $\ell(s) = s^3(1-s)^9$, which is obtained from the probability of three successes and nine failures in 12 trials.

A uniform prior will be assumed for s, so $f(s) = 1$ for all possible values of s between zero and one.

The simplest proposal distribution with which to generate possible values of the Markov chain is the uniform. Note that this need not be the same distribution as the prior; it is simply a coincidence in this example.

We can initiate the Markov chain with an arbitrary value $X_0 = 0.5$, and generate the next possible value by making a random draw from a uniform distribution. Assume that this next possible value is 0.45. To decide whether to accept 0.45 as the next value of the Markov chain we evaluate $f(X_0)\ell(X_0)$ and $f(Y)\ell(Y)$, which are equal to 0.5^{12} (0.000244) and $0.45^3 0.55^9$ (0.000420) respectively. Because the ratio of these two numbers is greater than one, Y (0.45) is accepted as the next value of the Markov chain.

The process is repeated with a new value of Y being generated by drawing it randomly from a uniform distribution. Assume that this next value is 0.6. Then $f(X_1)\ell(X_1)$ and $f(Y)\ell(Y)$ are calculated with values $0.45^3 0.55^9$ (0.000420) and $0.6^3 0.4^9$ (0.000057) respectively. The ratio of these two numbers is 0.135, so Y is accepted as the next value of the Markov chain with probability 0.135. The decision about whether to accept Y is achieved by generating another uniform random. If that number is less than 0.135, then Y is accepted as X_2, but if not then X_1 is accepted as X_2. This process continues to be repeated until a sufficiently large sample is obtained. The posterior probability density function of s can be estimated from the values of X_t.

The Metropolis algorithm always accepts the value Y if the probability density at Y is greater than at X_t. If it is less, the probability of moving to that value is proportional to its density. Hence, the algorithm is more likely to generate samples from parts of the posterior distribution that have high density.

Using this style of algorithm has two main consequences. Firstly, X_{t+1} typically depends on X_t. Any dependence means that each new sample provides a fraction of the information about the posterior distribution compared to an uncorrelated sample. An extremely large number of samples will be needed to obtain a good estimate of the posterior distribution if the correlation is particularly strong.

The second main consequence is that the initial part of the Markov chain is influenced by the arbitrary starting value, so the first part of the chain needs to be discarded as a 'burn-in' until the influence of the initial value is no longer apparent.

The proposal distribution can be chosen to optimize the performance of the Markov chain. For example, if the values of Y are very similar to X_t, then the value of R will be close to one and new values will be chosen in most iterations of the algorithm. However, the correlation of those samples will be high because $X_{t+1} \approx X_t$. In contrast, if the proposal allows large moves, then many of the new possible values will be far from the middle of the posterior distribution, leading to low values of R and very few instances in which Y is accepted as the new values of the chain. Therefore, when there are changes X_{t+1} and X_t will be very different but in most cases $X_{t+1} = X_t$. This also leads to strong correlation between successive samples of the chain. A happy medium is obtained when the possible values Y are not too far from X_t, which can be determined from trials runs of the algorithm during which it is adapted for optimal performance.

Why does it work?

Given a current value X_t, define the conditional probability of sampling X_{t+1} from Metropolis's Markov chain as $\Pr(X_{t+1} \mid X_t)$, and $g(X_{t+1})$ as the independent probability of sampling X_{t+1}. Proving that the samples from the Markov chain are samples from the posterior density $p()$ amounts to proving that $g()$ and $p()$ are equivalent probability density functions.

Bayes' rule provides:

$$\Pr(X_t \mid X_{t+1}) \times g(X_{t+1}) = \Pr(X_{t+1} \mid X_t) \times g(X_t), \text{ so}$$

$$g(X_t) = \Pr(X_t \mid X_t + 1) \times g(X_{t+1}) / \Pr(X_{t+1} \mid X_t).$$

The conditional probability $\Pr(X_{t+1} \mid X_t)$ is equal to the probability of the proposal being equal to X_{t+1} (given by $q(X_{t+1}, X_t)$) multiplied by the probability of the proposal being accepted. Assume that $p(X_{t+1}) < p(X_t)$, so the probability of acceptance is $p(X_{t+1})/p(X_t)$. Thus, $\Pr(X_{t+1} \mid X_t) = q(X_{t+1}, X_t) \times p(X_{t+1})/p(X_t)$.

The other conditional probability $\Pr(X_t \mid X_{t+1})$ is equal to the probability of the proposal moving from X_{t+1} to X_t ($q(X_t, X_{t+1})$) multiplied by the probability of the proposal being accepted. Because $p(X_{t+1}) < p(X_t)$, the acceptance probability is equal to one, so $\Pr(X_t \mid X_{t+1}) = q(X_t, X_{t+1})$. Thus,

$$g(X_t) = q(X_t, X_{t+1}) \times g(X_{t+1}) / [q(X_{t+1}, X_t) \times p(X_{t+1})/p(X_t)].$$

Because $q(X_t, X_{t+1}) = q(X_{t+1}, X_t)$ for the Metropolis algorithm (due to the symmetry of the proposal distribution), the terms for $q()$ cancel and

$$g(X_t) = g(X_{t+1}) / [p(X_{t+1})/p(X_t)], \text{ which leads to}$$

$$p(X_{t+1})/p(X_t) = g(X_{t+1})/g(X_t)$$

This equation demonstrates that the ratio of any two values of the probability density function $g()$ is the same as $p()$. Therefore, $p()$ and $g()$ are exactly proportional to each other. Also, because they are both probability density functions, $p()$ and $g()$ have the same area under the curve (which is equal to one), meaning that they must be identical functions not just in the same proportion.

The above proof used the assumption that $p(X_{t+1}) < p(X_t)$. One can reverse this assumption to give $p(X_{t+1}) > p(X_t)$, which leads to

$\Pr(X_{t+1} \mid X_t) = q(X_{t+1}, X_t)$ and $\Pr(X_t \mid X_{t+1}) = q(X_t, X_{t+1}) \times p(X_t)/p(X_{t+1})$. This then leads to:

$$g(X_t) = [p(X_t)/p(X_{t+1})] \times g(X_{t+1}), \text{ and again}$$

$$p(X_{t+1})/p(X_t) = g(X_{t+1})/g(X_t).$$

So the proof that $p() = g()$ does not depend on the assumption about the relative magnitude of $p(X_{t+1})$ and $p(X_t)$. Chib and Greenberg (1995) provide a more detailed discussion of the Metropolis-Hastings algorithm.

References

Adcock, C. J. (1997). Sample size determination: a review. *The Statistician*, **46**, 261–83.

Agresti, A. (1990). *Categorical Data Analysis.* New York, USA: Wiley.

Akaike, H. (1973). Information theory as an extension of the maximum likelihood principle. *Second International Symposium on Information Theory*, ed. B. N. Petrov and F. Caski. Budapest: Akademiai Kiado, pp. 267–81.

Akçakaya, H. R. (1990). A method for simulating demographic stochasticity. *Ecological Modelling*, **54**, 133–36.

Albert, J. H. (1997). Bayesian testing and estimation of association in a two-way contingency table. *Journal of the American Statistical Association*, **92**, 685–93.

Alefeld, G. and Herzberger, J. (1983). *Introduction to Interval Computations.* New York, USA: Academic Press.

Anderson, D. R., Burnham, K. P. and Thompson, W. L. (2000). Null hypothesis testing: problems, prevalence and an alternative. *Journal of Wildlife Management*, **64**, 912–23.

Anderson, J. L., (1998). Embracing uncertainty: the interface of Bayesian statistics and cognitive psychology. *Conservation Ecology 2*, http://www.ecologyandsociety.org/vol2/iss1/art2/

Arnquist, G. and Wooster, D. (1995). Meta-analysis: synthesizing research findings in ecology and evolution. *Trends in Ecology and Evolution*, **10**, 236–40.

Attiwill, P. M. and Leeper, G. W. (1987). *Forest Soils and Nutrient Cycles.* Carlton, Australia: Melbourne University Press.

Ayton, P. and Wright, G. (1994). Subjective probability: what should we believe? In *Subjective Probability*, ed. G. Wright and P. Ayton. New York, USA: Wiley, pp. 163–84.

Ayyub, B. M. (2001). *Elicitation of Expert Opinions for Uncertainty and Risks.* Boca Raton, USA: CRC Press.

Bakan, D. (1966). The test of significance in psychological research. *Psychological Bulletin*, **66**, 423–37.

Balakrishnan, N. and Nevzorov, V. B. (2003). *A Primer on Statistical Distributions.* Hoboken, NJ, USA: Wiley.

Bayes, T. R. (1763). An essay towards solving a problem in the doctrine of chances. *Philosophical Transactions,* **53**, 370−418.

Begon, M., Townsend, C. and Harper, J. (2005). *Ecology: From Individuals to Ecosystems,* 4th edn. Malden, MA, USA: Blackwell.

Belia, S., Fidler, F., Williams, F. and Cumming G. (2005). Researchers misunderstand confidence intervals and standard error bars. *Psychological Methods,* **10**, 389−96.

Bellhouse, D. R. (2004). The Reverend Thomas Bayes, FRS: A Biography to Celebrate the Tercentenary of His Birth. *Statistical Science,* **19**, 3−43.

Ben-Haim, Y. (2001). *Information-gap Decision Theory: Decisions Under Severe Uncertainty.* San Diego, USA: Academic Press.

Berger, J. O. (1985). *Statistical Decision Theory and Bayesian Analysis.* New York, USA: Springer-Verlag.

Berger, J. O. and Sellke, T. (1987). Testing a point null hypothesis: the irreconcilability of P values and evidence. *Journal of the American Statistical Association,* **82**, 112−22.

Berger, J. O. and Berry, D. A. (1988). Statistical analysis and the illusion of objectivity. *American Scientist,* **76**, 159−65.

Bilodeau, M. and Brenner, D. (1999). *Theory of Multivariate Statistics.* New York: Springer-Verlag.

Bondi, H. (2004). Correspondence: Statistics don't support cot-death murder theory: Misunderstanding of statistics is widespread and has led to miscarriages of justice. *Nature,* **428**, 799.

Bormann, F. H. and Likens, G. E. (1979). *Pattern and Process in a Forested Ecosystem.* New York, USA: Springer-Verlag.

Brack, C. L. (2002). Pollution mitigation and carbon sequestration by an urban forest. *Environmental Pollution,* **116**, S195−S200.

Brereton, R., Mallick, S. A. and Kennedy, S. J. (2004). Foraging preferences of swift parrots on Tasmanian blue-gum: tree size, flowering frequency and flowering intensity. *Emu,* **104**, 377−83.

Brooks, S. P. and Gelman, A. (1998). General methods for monitoring convergence of iterative simulations. *Journal of Computational and Graphical Statistics,* **7**, 434−55.

Broome, L. S. and Geiser, F. (1995). Hibernation in free-living Mountain Pygmy-possums, *Burramys parvus* (Marsupialia: Burramyidae). *Australian Journal of Zoology,* **43**, 373−79.

Brühl, C. A., Mohamed, V. and Linsenmair, K. E. (1999). Altitudinal distribution of leaf litter ants along a transect in primary forests on Mount Kinabalu, Sabah, Malaysia. *Journal of Tropical Ecology,* **15**, 265−77.

Burgman, M. A., Ferson, S. and Akcakaya, H. R. (1993). *Risk Assessment in Conservation Biology.* London, UK: Chapman and Hall.

Burgman, M. (2005). *Risks and Decisions for Conservation and Environmental Management.* Cambridge, UK: Cambridge University Press.

Burley, N., Krantzberg, G. and Radman, P. (1982). Influence of colour-banding on the conspecific preferences of zebra finches. *Animal Behaviour*, **30**, 444–55.

Burnham, K. P. and Anderson, D. R. (2002). *Model Selection and Multi-Model Inference: a Practical Information Theoretic Approach*. New York, USA: Springer-Verlag.

Calder, W. A. (1984). *Size, Function, and Life History*. Cambridge, MA, USA: Harvard University Press.

Carlin, B. P. and Chib, S. (1995). Bayesian model choice via Markov chain Monte Carlo methods. *Journal of the Royal Statistical Society*, **B57**, 473–84.

Carver, R. P. (1978). The case against statistical testing. *Harvard Educational Review*, **48**, 378–99.

Chambers, J. M., Cleveland, W. S., Kleiner, B. and Tukey, P. A. (1983). *Graphical Methods for Data Analysis*, CA, USA: Wadsworth.

Christensen, D. L., Herwig, B. R., Schindler, D. E. and Carpenter, S. R. (1996). Impacts of lakeshore residential development on coarse woody debris in north temperate lakes. *Ecological Applications*, **6**, 1143–49.

Chib, S. and Greenberg, E. (1995). Understanding the Metropolis-Hastings algorithm. *The American Statistican*, **49**, 327–35.

Clark, C. A. (1963). Hypothesis testing in relation to statistical methodology. *Review of Educational Research*, **33**, 455–73.

Clark, J. S. (2005). Why environmental scientists are becoming Bayesians. *Ecology Letters*, **8**, 2–15.

Clarke, R. D. (1972). The effect of toe clipping on survival in Fowler's toad (*Bufo woodhousei fowleri*). *Copeia*, 1972, 182–85.

Cohen, J. (1994). The earth is round (p < .05). *American Psychologist*, **49**, 997–1003.

Congdon, P. (2003). *Applied Bayesian Modelling*. Chichester, UK: Wiley.

Cox, R. T. (1946). Probability, frequency and reasonable expectation. *American Journal of Physics*, **14**, 1–13.

Crome, F. H. J., Thomas, M. R. and Moore, L. A. (1996). A novel Bayesian approach to assessing impacts of rain forest logging. *Ecological Applications*, **6**, 1104–23.

Dale, A. I. (1999) 2nd edn. *A History of Inverse Probability from Thomas Bayes to Karl Pearson*. New York: Springer.

Deming, W. E. (1975). On probability as a basis for action. *American Statistician*, **29**, 146–52.

Dennis, B. (1996). Discussion: should ecologists become Bayesians? *Ecological Applications*, **6**, 1095–1103.

Draper, D. (1995). Assessment and propagation of model uncertainty (with discussion). *Journal of the Royal Statistical Society*, **B57**, 45–97.

Dunning, J. B., Jr (1993). *CRC Handbook of Avian Body Masses*. Boca Raton, FL, USA: CRC Press.

Edwards, A. W. F. (1992). *Likelihood: an Account of the Statistical Concept of Likelihood and its Application to Scientific Inference*. Cambridge, UK: Cambridge University Press.

Elgar, M. A., Allan, R. A. and Evans, T. A. (1996). Foraging strategies in orb-spinning spiders: ambient light and silk decorations in *Agriope aetherea* Walckenaer (Araneae: Araneoidea). *Australian Journal of Ecology*, **21**, 464–67.

Elith, R. J. (2002). *Predicting the distribution of plants.* Ph.D. thesis, University of Melbourne, Parkville, Australia.

Ellison, A. M. (1996). An Introduction to Bayesian inference for ecological research and environment decision-making. *Ecological Applications*, **6**, 1036–46.

Ellison, A. M. (2001). Exploratory data analysis and graphical display. In *Design and Analysis of Ecological Experiments*, 2nd edn, ed. S. M. Scheiner and J. Gurevitch. Oxford: Oxford University Press, pp. 37–62.

Ellison, A. M. (2004). Bayesian inference in ecology. *Ecology Letters*, **7**, 509–20.

Ferson, S. (2005). *Bayesian Methods in Risk Assessment.* http://www.ramas.com/bayes.pdf

Ferson, S. (2002). *RAMAS Risk Calc 4.0 Software: Risk Assessment with Uncertain Numbers.* Boca Raton, USA: Lewis Publishers.

Fidler, F. (2005). *From Statistical Significance to Effect Estimation: Statistical Reform in Psychology, Medicine and Ecology.* Ph.D. thesis, University of Melbourne, Parkville, Australia.

Fidler, F., Cumming, G., Burgman, M. and Thomason, N. (2004). Statistical reform in medicine, psychology and ecology. *The Journal of Socio-Economics*, **33**, 615–30.

Fidler, F., Burgman, M. A., Cumming, G., Buttrose, R. and Thomason, N. (2006). Impact of criticism of null hypothesis significance testing on statistical reporting practices in conservation biology. *Conservation Biology*, **20**, 1539–44.

Fisher, R. F. (1930). Inverse probability. *Proceedings of the Cambridge Philosophical Society*, **26**, 528–35.

Flueck, W. T. (2001). Offspring sex ratio of introduced red deer in Patagonia, Argentina after an intensive drought. *Journal of Neotropical Mammalogy*, **8**, 139–47.

Forrester, G. E. and Steele, M. A. (2004). Predators, prey refuges, and the spatial scaling of density-dependent prey mortality. *Ecology*, **85**, 1332–42.

Fowler, J., Cohen, L. and Jarvis, P. (1998). *Practical Statistics for Field Biology*, 2nd edn. Chichester, UK: Wiley.

French, K. and Westoby, M. (1996) Vertebrate-dispersed species in a fire-prone environment. *Australian Journal of Ecology*, **21**, 379–85.

Gauthier-Clerc, M., Gendner, J.-P., Ribic, C. A., Fraser, W. R., Woehler, E. J., Descamps, S., Gilly, C., Le Bohec, C. and Le Maho, Y. (2004). Long-term effects of flipper bands on penguins. *Proceedings of the Royal Society of London*, **B271**, S423–26.

Gelman, A. and Meng, X.-L. (1996). Model checking and model improvement. In *Markov Chain Monte Carlo in Practice*, ed. W. R. Gilks, S. Richardson and D. J. Spiegelhalter. London, UK: Chapman and Hall, pp. 189–201.

Gelman. A., Carlin, J. B., Stern, H. S. and Rubin, D. B. (2004). *Bayesian Data Analysis*, 2nd edn. Boca Raton, FL, USA: Chapman and Hall/CRC.

Gibbons, P. and Lindenmayer, D. B. (2002). *Tree Hollows and Wildlife Conservation in Australia*. Melbourne, Australia: CSIRO Publishing.

Gigerenzer, G. and Hoffrage, U. (1995). How to improve Bayesian reasoning without instruction: frequency formats. *Psychological Review*, **102** (4), 684−704.

Gilks, W. R., Richardson, S. and Spiegelhalter, D. J. (1996). *Markov Chain Monte Carlo in Practice*. London, UK: Chapman and Hall.

Gilpin, M. E. and Soulé, M. E. (1986). Minimum viable populations: processes of species extinctions. *Conservation Biology: the Science of Scarcity and Diversity*, ed. M. E. Soulé. Sunderland, MA, USA: Sinauer, pp. 19−34.

Ginzburg, L. R., Slobodkin, L. B., Johnson, K. and Bindman, A. G. (1982). Quasiextinction probabilities as a measure of impact on population growth. *Risk Analysis*, **2**, 171−81.

Gotelli, N. J. and Arnett, A. E. (2000). Biogeographic effects of red fire ant invasion. *Ecology Letters*, **3**, 257−61.

Gotelli, N. J. and Ellison, A. M. (2004). *A Primer of Ecological Statistics*. Sunderland, MA, USA: Sinauer.

Grand, J. B., Flint, P. L., Peterson, M. R. and Moran, C. L. (1998). Effect of lead poisoning on spectacled eider survival rates. *Journal of Wildlife Management*, **62**, 1103−9.

Green, P. T. (1997). Red crabs in rain forest on Christmas Island, Indian Ocean: activity patterns, density and biomass. *Journal of Tropical Ecology*, **13**, 17−38.

Gurevitch, J. and Hedges, L. V. (2001). Meta-analysis: combining the results of independent experiments. *Design and Analysis of Ecological Experiments*, eds. S. M. Scheiner and J. Gurevitch, 2nd edn. Oxford, UK: Oxford University Press, pp. 347−69.

Haller, H. and Krauss, S. (2002). Misinterpretations of significance: a problem students share with their teachers? *Methods of Psychological Research Online*, **7** (1), pp. 1−20.

Hampton, J. M., Moore, P. G. and Thomas, H. (1973). Subjective probability and its measurement. *Journal of the Royal Statistical Society*, Series A, **136**, 21−42.

Harper, M. J., McCarthy, M. A. and van der Ree, R. (2005). The abundance of hollow-bearing trees in urban dry sclerophyll forest and the effect of wind on hollow development. *Biological Conservation*, **122**, 181−92.

Hastings, W. K. (1970). Monte Carlo sampling methods using Markov chains and their applications. *Biometrika*, **57**, 97−109.

Hero, J.- M. (1989). A simple code for toe clipping anurans. *Herpetological Review*, **20**, 66−7.

Hilborn, R. and Mangel, M. (1997). *The Ecological Detective: Confronting Models with Data*. Princeton, NJ, USA: Princeton University Press.

Hill, R. (2004). Multiple sudden infant deaths−coincidence or beyond coincidence? *Paediatric and Perinatal Epidemiology*, **18**, 320−26.

Hoenig J. M. and Heisey D. M. (2001). The abuse of power: the pervasive fallacy of power calculations for data analysis. *The American Statistician*, **55**, 19−24.

Hoeting, J. A., Madigan, D., Raftery, A. E. and Volinsky, C. T. (1999). Bayesian model averaging: a tutorial. *Statistical Science*, **14**, 382–401.

Howson, C. and Urbach, P. (1991). Bayesian reasoning in science. *Nature*, **350**, 371–4.

Huang, Y. J. (1987). The potential of vegetation in reducing summer cooling loads in residential buildings. *Journal of Climate and Applied Meteorology*, **26**, 1103–16.

Humphries, R. B. (1979). *Dynamics of a Breeding Frog Community*. Ph.D. thesis. The Australian National University.

Hunt, S., Cuthill, I. C., Swaddle, J. P. and Bennett, A. T. D. (1997). Ultraviolet vision and band-colour preferences in female zebra finches, *Taeniopygia guttata*. *Animal Behaviour*, **54**, 1383–92.

Jaynes, E. T. (1976). Confidence intervals vs. Bayesian intervals. *Foundations of Probability Theory, Statistical Inference, and Statistical Theories of Science*, II, eds. W. L. Harper and C. A. Hooker. Dordrecht, Holland: Reidel, pp. 175–213.

Jaynes, E. T. (2003). *Probability Theory: The Logic of Science*. New York, USA: Cambridge University Press.

Jeffreys, H. (1961). *Theory of Probability*, 3rd edn. Oxford, UK: Oxford University Press.

Johnson, N. L., Kotz, S. and Balakrishnan, N. (1994). *Continuous Univariate Distributions*, **1**, 2nd edn. New York, USA: Wiley.

Johnson, N. L., Kotz, S. and Balakrishnan, N. (1995). *Continuous Univariate Distributions*, **2**, 2nd edn. New York, USA: Wiley.

Johnson, N. L., Kotz, S. and Balakrishnan, N. (1997). *Discrete Multivariate Distributions*, 2nd edn. New York, USA: Wiley.

Johnson, N. L., Kotz, S. and Kemp, A. W. (1992). *Univariate Discrete Distributions*, 2nd edn. New York, USA: Wiley.

Johnson, D. H. (1995). Statistical sirens: the allure of nonparametrics. *Ecology*, **76**, 1998–2000.

Johnson, D. H. (1999). The insignificance of statistical significance testing. *Journal of Wildlife Management*, **63**, 763–72.

Johnston, J. P., Peach, W. J., Gregory, R. D. and White, S. A. (1997). Survival rates of tropical and temperate passerines: a Trinidadian perspective. *American Naturalist*, **150**, 771–89.

Joyce, H. (2002). Beyond reasonable doubt. *Plus Magazine*, **21** (http://pass.maths.org.uk/issue21/features/clark/index.html)

Kahneman, D., Slovic, P. and Tversky A. (eds.) (1982). *Judgement Under Uncertainty: Heuristics and Biases*. New York: Cambridge University Press.

Kass, R. E. and Raftery, A. E. (1995). Bayes factors and model uncertainty. *Journal of the American Statistical Association*, **90**, 773–95.

Kaufmann, A. and Gupta, M. M. (1985). *Introduction to Fuzzy Arithmetic*. New York, USA: Reinhold.

Knuth, D. E. (1997). *The Art of Computer Programming. Semi-numerical Algorithms*, **2**, 3rd edn. Reading, MA, USA: Addison-Wesley.

Körtner, G. and Geiser, F. (1998). Ecology of natural hibernation in the marsupial mountain pygmy-possum (*Burramys parvus*). *Oecologia*, **113**, 170–78.

Kotz, S., Balakrishnan, N. and Johnson, N. L. (2000). *Continuous Multivariate Distributions*, 2nd edn. New York, USA: Wiley.

Lebreton, J.-D., Burnham, K. P., Clobert, J. and Anderson, D. R. (1992). Modeling survival and testing biological hypotheses using marked animals: a unified approach with case studies. *Ecological Monographs*, **62**, 67–118.

Lemckert, F. (1996). Effects of toe-clipping on the survival and behaviour of the Australian frog *Crinia signifera*. *Amphibia-Reptilia*, **17**, 287–90.

Lindley, D. V. (1997). The choice of sample size. *The Statistician*, **46**, 129–38.

Lindley, D. V. and Phillips, L. D. (1976). Inference for a Bernoulli process (a Bayesian view). *The American Statistician*, **30**, 112–19.

Link, W. A. and Barker, R. J. (2006). Model weights and the foundation of multimodel inference. *Ecology*, **87**, 2626–35.

Lüddecke, H. and Amézquita, A. (1999). Assessment of disc clipping on the survival and behaviour of the Andean frog *Hyla labialis*. *Copeia*, 1999, 824–30.

Ludwig, D. (1996). Uncertainty and the assessment of extinction probabilities. *Ecological Applications*, **6**, 1067–76.

Mackenzie, D. I., Nichols, J. D., Lachman, G. B., Droege, S., Royle, J. A. and Langitimm, C. A. (2002). Estimating site occupancy rates when detection probabilities are less than one. *Ecology*, **83**, 2248–55.

Mansergh, I., Baxter, B., Scotts, D., Brady, T. and Jolley, D. (1990). Diet of *Burramys parvus* (Marsupialia: Burramyidae) and other small mammals in the alpine environment at Mt Higginbotham, Victoria. *Australian Mammalogist*, **13**, 167–77.

Mansergh, I. M. and Broome, L. S. (1994). *The Mountain Pygmy-possum of the Australian Alps*. Sydney, Australia: University of New South Wales Press.

Martin, T. G., Kuhnert, P. M., Mengersen, K. and Possingham, H. P. (2005). The power of expert opinion in ecological models using Bayesian methods: impact of grazing on birds. *Ecological Applications*, **15**, 266–80.

Marzolin, G. (1988). Polygynie du Cincle pongeur (*Cinclus cinclus*) dans les côtes de Lorraine. *L'Oiseau et la Revue Francaise d'Ornithologie*, **58**, 277–86.

Masters, P. (1993). The effects of fire-driven succession and rainfall on small mammals in spinifex grasslands at Uluru National Park, Northern Territory. *Wildlife Research*, **20**, 803–13.

Masters, P., Dickman, C. and Crowther, M. (2003). The effects of cover reduction on Mulgara (*Dasycercus cristicauda*), rodent and invertebrate populations in central Australia: implications for management. *Austral Ecology*, **28**, 658–65.

May, R. M. (2004). Ethics and amphibians. *Nature*, **431**, 403.

McCarthy, M. A. (1996). Red kangaroo (*Macropus rufus*) dynamics: effects of rainfall, harvesting, density dependence and environmental stochasticity. *Journal of Applied Ecology*, **33**, 45–53.

McCarthy, M. A. (1997). Competition and dispersal from multiple nests. *Ecology*, **78**, 873–83.

McCarthy, M. A. and Parris, K. M. (2004). Clarifying the effect of toe clipping on frogs with Bayesian statistics. *Journal of Applied Ecology*, **41**, 780–86.

McCarthy, M. A. and Broome, L. S. (2000). A method for validating stochastic models of population viability: a case study of the mountain pygmy-possum (*Burramys parvus*). *Journal of Animal Ecology*, **69**, 599–607.

McCarthy, M. A. and Thompson, C. (2001). Expected minimum population size as a measure of threat. *Animal Conservation*, **4**, 351–55.

McCarthy, M. A. and Masters, P. (2005). Profiting from prior information in Bayesian analyses of ecological data. *Journal of Applied Ecology*, **42**, 1012–19.

McCarthy, M. A., Franklin, D. C. and Burgman, M. A. (1994). The importance of demographic uncertainty: an example from the helmeted honeyeater. *Biological Conservation*, **67**, 135–42.

McCarthy, M. A., Webster, A., Loyn, R. H. and Lowe, K. W. (1999). Uncertainty in assessing the viability of the powerful owl *Ninox strenua* in Victoria, Australia. *Pacific Conservation Biology*, **5**, 144–54.

McCullagh, P. and Nelder, J. A. (1989). *Generalized Linear Models*, 2nd edn, London, UK: Chapman and Hall.

McLean, N. (2003). *Ecology and Management of Overabundant Koala (Phascolarctos cinereus) Populations*. Ph.D. thesis. University of Melbourne, Parkville, Australia.

McPherson, E. G., Scott, K. I. and Simpson, J. R. (1998). Estimating cost effectiveness of residential yard trees for improving air quality in Sacramento, California, using existing models. *Atmospheric Environment*, **32**, 75–84.

Metropolis H., Rosenbluth A. W., Rosenbluth M. N., Teller A. H. and Teller E. (1953). Equations of state calculations by fast computing machines. *Journal of Chemical Physics*, **21**, 1087.

Morgan, M. G. and Henrion, M. (1990). *Uncertainty: a Guide to Dealing with Uncertainty in Quantitative Risk and Policy Analysis*. Cambridge, UK: Cambridge University Press.

Mullan Crain, C., Silliman, B. R., Bertness, S. L. and Bertness, M. D. (2004). Physical and biotic drivers of plant distribution across estuarine salinity gradients. *Ecology*, **85**, 2539–49.

O'Donnell, T. (1936). *History of Life Insurance in Its Formative Years; Compiled from Approved Sources*. Chicago, USA: American Conservation Company.

O'Hagan, A. and Luce, B. R. (2003). A primer on Bayesian statistics in health economics and outcomes research. *Bayesian Initiative in Health Economics & Outcomes Research*. Bethesda, Maryland: Bayesian Initiative in Health Economics and Outcomes Research; Sheffield, UK: The Centre for Bayesian Statistics in Health Economics.

Oakes, M. (1986). *Statistical Inference: A Commentary for the Social and Behavioural Sciences*. Chichester, UK: John Wiley & Sons Ltd.

Parkhurst, D. F. (1997). *Commentaries on Significance Testing*. http://www.indiana.edu/~stigtsts/index.html#contents

Parris, K. M. (2001). Distribution, habitat requirements and conservation of the cascade treefrog (*Litoria pearsoniana*, Anura: Hylidae). *Biological Conservation*, **99**, 285–92.

Parris, K. M. and McCarthy, M. A. (2001). Identifying effects of toe-clipping on anuran return rates: the importance of statistical power. *Amphibia-Reptilia*, **22**, 275–89.

Parris, K. M. (2006). Urban amphibian assemblages as meacommunities. *Journal of Animal Ecology*, **75**, 757–64.

Paruelo, J. M. and Laueroth, W. K. (1996). Relative abundance of plant functional types in grasslands and shrublands of North America. *Ecological Applications*, **6**, 1212–24.

Peterman, R. M. (1990). Statistical power analysis can improve fisheries research and management. *Canadian Journal of Aquatic Sciences*, **47**, 2–15.

Peters, R. H. (1983). *The Ecological Implications of Body Size*. Cambridge, UK: Cambridge University Press.

Polis, G. A., Hurd, S. D., Jackson, C. D. and Sanchez-Piñero, F. (1998). Multifactor population limitation: variable spatial and temporal control of spiders on Gulf of California islands. *Ecology*, **79**, 490–502.

Press, W. H., Teukolsky, S. A., Vetterling, W. T. and Flannery, B. P. (1992). *Numerical Recipes in C: The Art of Scientific Computing*. Cambridge, UK: Cambridge University Press.

Quinn, G. P. and Keough, M. J. (2002). *Experimental Design and Data Analysis*. Cambridge, UK: Cambridge University Press.

Richards, S. A. (2005). Testing ecological theory using the information-theoretic approach: examples and cautionary results. *Ecology*, **86**, 2805–14.

Rozeboom, W. W. (1997). Good science is abductive, not hypothetico-deductive. In *What If There Were No Significance Tests?*, ed. L. L. Harlow, S. A. Mulaik and J. H. Steiger. Hillsdale, NJ, USA: Erlbaum, pp. 335–92.

Savage, V. M., Gillooly, J. F., Brown, J. H., West, G. B. and Charnov, E. L. (2004). Effects of body size and temperature on population growth. *American Naturalist*, **163**, 429–41.

Shaffer, M. L. (1981). Minimum population sizes for species conservation. *Bioscience*, **31**, 131–4.

Smith, A. and Broome, L. S. (1992). The effects of environment and sex on the diet of the Mountain Pygmy-possum and its implications for the species' conservation and management in south-east Australia. *Australian Wildlife Research*, **19**, 755–68.

Sokal, R. R. and Rohlf, F. J. (1995). *Biometry: The Principles and Practice of Statistics in Biological Research*, 3rd edn. New York: W. H. Freeman and Co.

Spiegelhalter, D., Thomas, A., Best, N. and Lunn, D. (2005). *WinBUGS User Manual Version 2.10*. Cambridge, UK: MRC Biostatistics Unit.

Spiegelhalter, D. J, Best, N. G., Carlin, B. P. and van der Linde, A. (2002). Bayesian measures of model complexity and fit. *Journal of the Royal Statistical Society: Series B*, **64**, 583–639.

Stephens, P. A., Buskirk, S. W., Hayward, G. D. and Martínez Del Rio, C. (2005). Information theory and hypothesis testing: a call for pluralism. *Journal of Applied Ecology*, **42**, 4–12.

Stow, C. A. and Borsuk, M. E. (2003). Enhancing causal assessment of estuarine fishkills using graphical models. *Ecosystems*, **6**, 11–19.

Taylor, B. L. and Gerrodette, T. (1993). The uses of statistical power in conservation biology: The vaquita and Northern Spotted Owl. *Conservation Biology*, **7**, 489–500.

Trivers, R. L. and Willard, D. E. (1973). Natural selection of parental ability to vary the sex ratio of offspring. *Science*, **179**, 90–1.

Tukey, J. W. (1997). *Exploratory data analysis*. Reading, MA, USA: Addison-Wesley.

Tversky, A. and Kahneman, D. (1974). Judgment under uncertainty: heuristics and biases. *Science*, **185**, 1124–31.

Tyre, A. J., Tenhumberg, B., Field, S., Possingham, H. P., Niejalke, D. and Parris, K. (2003). Improving precision and reducing bias in biological surveys by estimating false negative error rates in presence-absence data. *Ecological Applications*, **13**, 1790–1801.

Underwood, A. J. (1997). *Experiments in Ecology: Their Logical Design and Interpretation Using Analysis of Variance*. Cambridge, UK: Cambridge University Press.

Volinsky, C. T., Madigan, D., Raftery, A. E. and Kronmal, R. A. (1997). Bayesian model averaging in proportional hazard models: predicting the risk of a stroke. *Applied Statistics*, **46**, 433–48.

Wade, P. R. (2000). Bayesian methods in conservation biology. *Conservation Biology*, **14**, 1308–16.

Waichman, A. V. (1992). An alphanumeric code for toe clipping amphibians and reptiles. *Herpetological Review*, **23**, 19–21.

Walley, P. (1991). *Statistical Reasoning with Imprecise Probabilities*. London, UK: Chapman and Hall.

West, B. G., James, H., Brown, J. H. and Enquist, B. J. (1997). A general model for the origin of allometric scaling laws in biology. *Science*, **276**, 122–6.

Williamson, I. and Bull, C. M. (1996). Population ecology of the Australian frog, *Crinia signifera*: adults and juveniles. *Wildlife Research*, **23**, 249–66.

Wintle, B. A., and Bardos, D. C., (in press). Modelling species habitat relationships with spatially autocorrelated observation data. *Ecological Applications*.

Wintle, B. A., McCarthy, M. A., Parris, K. M. and Burgman, M. A. (2004). Precision and bias of methods for estimating point survey detection probabilities. *Ecological Applications*, **14**, 703–12.

Wintle, B. A., Kavanagh, R. P., McCarthy, M. A. and Burgman, M. A. (2005a). Estimating and dealing with detectability in occupancy surveys for forest owls and arboreal marsupials. *Journal of Wildlife Management*, **69**, 905–17.

Wintle, B. A., Elith J. and Potts J. M. (2005b). Fauna habitat modelling and mapping: a review and case study in the Lower Hunter Central Coast region of NSW. *Austral Ecology*, **30**, 719–38.

Wintle, B. A., McCarthy, M. A., Volinsky, C. T. and Kavanagh, R. P. (2003). The use of Bayesian Model Averaging to better represent uncertainty in ecological models. *Conservation Biology*, **17**, 1579–90.

Zar, J. H. (1999). *Biostatistical analysis*. Upper Saddle River, NJ, USA: Prentice Hall.

Ziliak, S. and McCloskey, D. (2004). Size matters: the standard error of regressions in the American Economic Review. *Journal of Socio-economics*, **33**, 527–47.

Index

Page numbers in bold refer to tables; page numbers in italic refer to figures and boxes.